Music Technology Essentials

Music Technology Essentials provides an overview of the vocabulary, techniques, concepts, and devices used in contemporary music production and guides readers through the essential fundamentals of music technology so that they can create their own music productions at home.

This highly accessible book covers sound fundamentals and theory, as well as practical topics like hardware, software, MIDI, digital audio, synthesis, computer notation, and audio-visual applications, to equip the reader with the principles they need to achieve professional-sounding results. Each chapter is accompanied by real-life examples and exercises that can be applied to any digital audio workstation software, to put the lessons into practice. This book will also help readers evaluate their requirements for home music production while working within a sensible budget.

Music Technology Essentials is the ideal textbook for beginners inside and outside of the classroom, including those on music and music production courses, who wish to enter the world of music technology but are unsure where to start or what to purchase.

Andrew Maz is Chair of the Music Department and Lead for the Commercial Music at Cerritos College. Andrew also works as a freelance composer, songwriter, and producer and is the lead and primary reviewer for music technology courses for the California Community College system.

Music Technology Essentials

A Home Studio Guide

Andrew Maz

Routledge
Taylor & Francis Group
LONDON AND NEW YORK

Designed cover image: © Daniel Myjones/Shutterstock.com

First published 2024
by Routledge
4 Park Square, Milton Park, Abingdon, Oxon OX14 4RN

and by Routledge
605 Third Avenue, New York, NY 10158

Routledge is an imprint of the Taylor & Francis Group, an informa business

British Library Cataloguing-in-Publication Data
A catalogue record for this book is available from the British Library

Library of Congress Cataloging-in-Publication Data
Names: Maz, Andrew, author.
Title: Music technology essentials : a home studio guide / Andrew Maz.
Description: Abingdon, Oxon ; New York : Routledge, 2023. | Includes index. | Identifiers: LCCN 2023016120 (print) | LCCN 2023016121 (ebook) | ISBN
9781032384542 (paperback) | ISBN 9781032384573 (hardback) |
ISBN 9781003345138 (ebook)
Subjects: LCSH: Sound recordings—Production and direction. | Popular music—Production and direction. | Digital audio editors. | Computer sound processing. | Software synthesizers. | Software samplers. | Software synthesizers. | Musical notation—Computer programs.
Classification: LCC ML3790 .M37 2023 (print) | LCC ML3790 (ebook) | DDC 781.49—dc23/eng/20230627
LC record available at https://lccn.loc.gov/2023016120
LC ebook record available at https://lccn.loc.gov/2023016121

ISBN: 978-1-032-38457-3 (hbk)
ISBN: 978-1-032-38454-2 (pbk)
ISBN: 978-1-003-34513-8 (ebk)

DOI: 10.4324/9781003345138

Typeset in Sabon
by codeMantra

Access the Companion Website: www.andrewmaz.net/essentials

Contents

Acknowledgments

I would like to thank my wife, Trina, for her patience and understanding during the writing of this book. I also wish to thank my sister, Elaine, who prepared all the original images for this book. I would like to acknowledge Jun Fujimoto (Yamaha), Sean Tokuyama (Steinberg), Javad Butah (Ableton), and Lee Whitmore (Focusrite) for their assistance in obtaining permission for the use of product images.

Finally, I would like to thank my students, whose questions and curiosity prompted the creation of this book.

Andrew Maz
www.andrewmaz.net

Introduction

Hello and welcome to *Music Technology Essentials*. This book grew out of the need to provide my students with a practical guide to introducing them to music technology. Students enter the realm of music technology with the goal of being a songwriter, producer, or composer. In my classes, I discovered that many of them made decisions on technology based on what they read and saw on the internet. I also learned that many of them lacked a basic understanding of concepts such as sound and digital audio. Their decisions are often based on the promises made by advertisements and artist endorsements. This often leads them to spend significant resources purchasing audio technology and software that often fails to fulfill those promises. The goal of this book is to educate the reader with essential information and understanding of music technology so that they can make informed decisions on the equipment they need.

The information my students gather on the internet helps them get up and running quickly to make music. However, they lack fundamental concepts about audio technology and software. They might know how to record audio and MIDI data, but they do not understand how this occurs. When they cannot accomplish more complicated tasks or do not see the results they want, they do not know how to proceed. They are sometimes led to believe that software and technology are the reasons for their troubles. This compels them to spend more money on technology promising to solve their problems.

This book is not a single-ended solution. Instead, it offers the reader a platform to begin with so that they gain confidence with technology and have the knowledge to increase their knowledge as their needs grow. Fundamental material is presented in practical terms, and the examples and exercises illustrate the concepts so that the reader fully comprehends the material. I sincerely hope that whether you are a student or not, you find the information in this book helpful and practical.

DOI: 10.4324/9781003345138-1

Chapter 1

The Home Studio

The Home Studio

The mention of a recording studio brings up images of large rooms with musicians gathered around microphones and a recording engineer sitting behind a large mixing console looking at the musicians through a window. We imagine expensive equipment and a collection of microphones and speakers designed to make music sound its best. Recording and mixing engineers seem like magicians who can somehow pull all the elements together and create a cohesive sound. The idea of being able to create professional-sounding music in any other space by yourself seemed impossible.

Twenty years ago, the ability to have a studio in your home to produce music of any quality was costly and difficult. Computer digital recording was a new technology, and it required specific hardware and computer configurations. Those systems would often record only two tracks at a time. You could record drums with two microphones. Then add a bass and guitar on the next pass. Pianos and synthesizers could be added after that and then the vocals. These were not ideal conditions, but at least it was digital.

Cassette recording systems were an option. The audio quality of these units was less than ideal, but cassettes were inexpensive, and you could record four tracks. If you need more tracks, you would need to "bounce" tracks to create new ones. You would record on the first three tracks, say drums, bass, and guitar. You would then create a balance between those three tracks so that you could hear all three instruments clearly and record them onto the fourth track. If the mix of the three instruments was satisfactory, you could erase the first three tracks and add more sounds. The problem with this system is that if later you decide that the bass should be louder, you have no way of isolating the bass because it was on the same track as the drums and guitar. Once again, less than ideal conditions, but at least we could create and record music without having to go into a studio.

Fortunately, today we have more options for music creation at home. Computers are significantly faster, and storage is plentiful. Audio software no longer requires specialized hardware and computers. This means we can

DOI: 10.4324/9781003345138-2

all record digitally. Audio software can run virtual synthesizers and acoustic instruments such as pianos, guitars, and drums. Audio interfaces accommodate multiple inputs allowing us to record more than two instruments at once. The quality of entry-level microphones and speakers has improved significantly, allowing us to record and playback audio with more clarity. The computer makes it possible to access studio-quality effects at home, without going into a studio. We have a range of options. But how do you choose the best solution for your needs? How do you know if the product delivers what it promises? How do you determine what equipment you need for your music production?

In this book, we will cover the concepts and technology associated with music production. We will cover the basics of sound, digital audio concepts, computers, audio hardware and software, and synthesizers. We will explore technology options for music creation. Our focus is to find cost-effective solutions to get you started with producing music at home. The type(s) of music you wish to create can dictate your technological requirements. You will learn how to assess your needs and use those requirements to determine technology purchases. You will understand the terminology associated with audio technology so that equipment specifications guide your decisions, not advertisements.

You probably are aware that manufacturers, advertisers, and reviewers present dozens of solutions for your audio production needs. Many of these presentations are convincing with fair pricing and promising results. However, given the amount of information available to you through the internet, it may be difficult to determine the accuracy of the information and if there is any personal bias from the reviewer. This can make decisions on what to buy confusing and costly. Over the years, I have purchased technology solutions with the hope of solving my production problems. I have hardware and software that goes unused. It is not that the solutions were inadequate, but they did not provide the results I needed to complete my work. Therefore, we need to do our research before we purchase hardware and software. But first, we need to determine our goals.

Goals

The goal of the home studio is to create a comfortable, creative, and productive environment for music production. The studio requires reliable equipment and a work environment that is not distracting. Your chosen audio application should integrate with your workflow and offer features that appeal to you. Your own workflow is important because we all create differently. The hardware and software you choose should allow you to work in the method that best suits you. Some audio applications require you to follow a specific order of events when creating, others allow for flexibility.

The same is true for the hardware we use. If an audio interface requires additional steps to connect and power a microphone because you must access the controls from another software application, then your workflow is interrupted.

Your home studio must be stable and reliable. Your equipment should always work when you need it to work. Faulty equipment is distracting and time-consuming. Imagine starting to create a mix of a song and the sound to your left speaker cuts in and out. This fault requires you to stop your work, diagnose, and solve the problem. This small issue halts your creative process until you restore the sound. Creativity is unpredictable and often fleeting. You do not want to wait for your computer to install updates when you want to begin working or interrupt you while you are working. To prevent such delays, you need to routinely check your equipment so that all connections are secure, and the computer is up to date.

It is best to work in an environment where all your equipment is connected and ready for work. Having to rearrange furniture or clear a desk and connect equipment every time you want to work takes time. If you do not have a dedicated workspace, then you need to configure a portable setup that can be quickly configured for work. Some of you may prefer this since you are free to work in any space you like. Others will prefer having a dedicated workspace and a dedicated computer strictly for music creation.

I am often asked by new users "what is the best set of speakers I can buy?" This question is difficult to answer because the value of audio technology is highly subjective, and opinions differ. A more appropriate question is "what are the best speakers I can afford?" The subtle distinction is that the question focuses on your needs and your available resources. However, even more important is determining what type of speakers you really need. This is a difficult question to answer if you are new to music technology, which is one of the reasons you might be reading this book. In Chapter 7, you will learn about speaker design, types, and placement. Understanding how speakers work will help you determine which speakers you need.

Your home studio can be more than a sketchpad for your musical ideas. A thought-out home studio can produce professional results, even with entry-level equipment. Understanding how the technology works and its limitations allows you to achieve quality results. Investing in expensive speakers is not a wise decision if your work environment is filled with background noise and interruptions. The traffic noise outside your window will prevent you from hearing all the details in the speakers. In this situation, you might benefit from a good pair of headphones.

Concepts

There are several concepts that we will discuss in this book. These concepts might seem insignificant at first but understanding these concepts will aid you in your journey with music production. This knowledge will help you communicate clearly with other musicians and guide you when evaluating audio hardware and software.

Technology

The ease with which we create our productions depends heavily on technology. Technology is expensive, so we want to choose products that provide us with good value for money. We can quickly spend money on technology, so understanding our immediate needs versus later needs can help us offset costs and develop budgets.

Sound

Chapter 2 covers the fundamentals of sound. We will learn how sound is created and moves through the air, interacting with our ears and audio equipment. We will learn how to describe sound and how to measure its properties. We will discover the properties of sound that allow instruments and voices to sound different to our ears. We will spend time learning about our hearing, the parts of our ears, how we process sound, and most importantly, how to protect our hearing from permanent damage. Our knowledge of sound and hearing will help us appreciate how digital audio works in our computers.

Digital Audio

The development of digital audio and the ability to process digital audio in a computer are what allow us to create our home studio. The most basic digital audio workstation (DAW) with an entry-level application and audio interface can produce high-quality audio equal to that in a professional recording studio. Knowing how audio is digitized, processed, and stored helps us understand the need for a stable and fast computer with ample storage space. The specifications for digital audio vary depending on the application. For example, the specifications for audio-only projects are different than that for video projects. The significance of streaming audio also requires that we understand the difference between compressed and uncompressed audio. Chapter 3 explains digital audio.

Computers

The computer is the center of your home studio. The computer is responsible for running DAW applications. The DAW application emulates the features and functions of a professional recording studio. The DAW application allows you to record, edit, and mix audio sources. Digitizing audio is task intensive, so the computer must be powerful enough to process the audio with minimal delay and signal degradation. The storage devices must be fast enough to accurately capture audio and stream that information for playback. If you intend to work with virtual instruments, software synthesizers that run within the DAW application, the computer needs sufficient memory to cache the sounds for playback. DAW applications display waveforms in real time, requiring a computer with adequate graphics processing to render the images in real time. Finally, you need the correct ports to connect audio devices and keyboard controllers. The computer requirements for DAW applications are significant, and in Chapter 4, we will look at those requirements.

Digital Audio Workstation

A DAW is a collection of components that allow you to record, edit, and produce music on a computer. A basic DAW consists of a computer, a DAW application, an audio interface, and speakers or headphones. If you wish to play virtual instruments included with the DAW application, then you need a music keyboard controller. In Chapter 5, we will examine the components of a DAW and then focus our attention on the DAW application. We will learn the features of DAW applications, how they function, and what is included with the application. This chapter will explain audio effects and virtual instruments. You can choose from different DAW applications depending on your needs and budget. DAW applications differ in their workflows and focus.

Audio Effects

Audio effects are an important part of audio production. It is also a dense topic as each effect type varies in complexity and operation, depending on the manufacturer. Chapter 6 provides you with a broad overview of audio effects. We will look at the effect families and the common effects in each family. We will discuss the basic parameters of each effect and common uses for those effects. This chapter will provide you with the information to distinguish audio effects and gain a basic understanding on how they operate and sound.

Audio Hardware

Chapter 7 covers the essential audio hardware needed for music production. Each of these hardware requirements is covered in detail.

Audio Interface

An audio interface allows you to record and playback audio from your computer. The audio interface converts acoustic sound into digital data that can be stored and retrieved from your computer. The audio interface determines the number of audio sources you can record simultaneously. An audio interface with two audio inputs limits you to record two sources at once. If you are a singer/songwriter, you have one for your voice and the other for a guitar. This might be perfect for your goals. If you intend to record a drum set, you will need more audio input connections. Electronic musicians working with virtual instruments may find two inputs more than enough. All audio interfaces offer at least a pair of outputs to connect speakers. They will also have at least one stereo headphone connection. However, if you plan to work with multiple musicians, then multiple headphone outputs are beneficial. Your workflow and music production goals determine the number of simultaneous audio inputs needed.

Microphones

Acoustic instruments, such as guitar, piano, or voice, require a microphone to capture the audio with an audio interface. Microphone options are extensive, ranging from affordable to extremely expensive. A more expensive microphone does not always mean a better sound. Some microphones are better suited at capturing instruments rather than voices. There are microphones designed to capture specific instruments, such as drums. Some microphones are more sensitive than others while some are designed for live performance use and can handle the rigors of concerts. Very few microphones can accomplish all these tasks, so there is a good chance that you will need more than one microphone, depending on what you plan to record.

MIDI Controllers

MIDI stands for Musical Instrument Digital Interface. MIDI is a protocol that transmits note data between computers and physical synthesizers. If you have a synthesizer with its own keyboard, then you can record the note data to your computer with MIDI, provided your computer has a MIDI interface. DAW applications come with virtual instruments. These instruments run within the DAW and offer a variety of sounds ranging from synthesizers to orchestral instruments to acoustic instruments. Virtual instruments allow you to work with orchestral sounds without needing to hire an orchestra and rent a studio. Virtual instruments allow you to add drums to your music without needing to hire a drummer and record drums. These instruments can increase your sound palette and allow you to explore new musical arrangements.

To play virtual instruments in real time, you need a MIDI controller. A MIDI controller has piano keys but does not produce any sound. MIDI controllers connect to your computer with a USB connection and allow you to play virtual instruments within the DAW. MIDI controllers come in a variety of sizes. Some feature full-size keys while others feature mini keys for portability. The type of MIDI controller you purchase depends on your keyboard-playing skills. An experienced piano player will want full-size keys that feel like a piano. A less-experienced musician might find mini keys suitable because these keyboards take up less space and can be easily carried. The type of device you purchase depends on your workflow and how you intend to create music.

Speakers and Headphones

You need to be able to listen to your music creations with a pair of speakers or headphones. Headphones vary in price and features, and purchasing headphones is subjective. Headphones are convenient because they offer you privacy and isolation from outside noises. If you live in a busy neighborhood, then headphones can help you focus on the music without distractions. Headphones allow you to work at all hours without disturbing neighbors and guests. You can carry headphones with you and work in different locations if you have a laptop. We will explore headphone design and learn which headphones are suitable for music and which ones are not.

Speakers, like headphones, come in various sizes, specifications, and prices. Large speakers allow you to listen to music at loud levels. However, when placed in a small room, large speakers lose their effectiveness because they must be spaced far enough away from each other to create a stereo field. Smaller speakers may not be as loud, but they take up less space on a desk and can be placed closer to your ears. Purchasing monitors is a subjective experience. There are several budget-conscious speakers that offer excellent results without great expense.

Chapter 7 covers the design and function of audio interfaces, microphones, headphones, speakers, and MIDI controllers. This information will help you choose the hardware devices that best work for you.

MIDI

The MIDI protocol has been around since the 1980s. It is still present in our DAW applications. If you plan on using external synthesizers and drum machines, then you need to understand the MIDI protocol, its functions, and its connections. If you plan on using virtual instruments on your computer, you will record the notes as MIDI data and then edit and process that information. DAW applications make MIDI transparent, but advanced editing and processing requires knowledge of MIDI. If you want to compose music with virtual instruments, then you will be working with MIDI data when you record and edit. Chapter 8 covers the basics of MIDI.

Synthesis and Sampling

Chapter 9 explores synthesis and sampling. The virtual synthesizer instruments included with a DAW application are based on different types of synthesis. Each type of synthesis produces a characteristic sound. If you want a thick and prominent bass sound, then an analog synthesizer is a good place to start. Bells and other metallic sounds are found in digital synthesizers. Understanding the different synthesis methods will not only help you choose the appropriate synthesizer for the sound you are looking for but also give you the knowledge to edit and program sounds. You will be able to shape sounds to fit your needs.

Sampling records acoustic instruments for playback on a keyboard. Instruments that sample allow you to add pianos, guitars, orchestral, and other acoustic instruments to your productions. Sampling is convenient because you do not need to find a piano to record; you can use the samples included with your DAW application. Sampled instruments are not necessarily designed to replace real acoustic instruments, but they are practical when you need to add strings to a production. We will look at how sampled instruments are created and managed. The quality of sampled instruments has improved significantly in recent years, and depending on the context, listeners may not be able to determine if you are using real instruments or samples.

Notation

In Chapter 10, we will look at computer notation software. This might seem unnecessary at first, but notation software could be very useful depending on the music you create. A film composer will eventually need to create sheet music for musicians to read from if the score is performed by an orchestra. The songwriter may discover that the vocalist wants a lead sheet to learn the songs. The songwriter might need to create chord charts for the band to rehearse from, which can save time when they are learning a new song. Not everyone will need notation software, but it is worth exploring, just in case.

Needs and Budget

Understanding your needs and budget will help you decide on equipment. The type of music you create and produce will influence your needs. For example, if you want to create electronic music, then a MIDI controller is a more immediate need than a microphone. A songwriter working with acoustic instruments might need different microphones to record voices and instruments. A film composer might need microphones and a MIDI controller to work with virtual and real instruments.

With respect to audio interfaces, an electronic musician writing instrumental music can easily work with an interface with two inputs and outputs. Another creator may need to interface with DJ equipment and a sound system, in which case, additional outputs are needed. A songwriter may want additional audio inputs to record multiple musicians at once. They may even want an interface with more than one pair of headphone outputs. The composer can start with two inputs and outputs, but they may also wish to record multiple instruments at once, requiring an interface with multiple inputs. Some instruments require multiple microphones, such as a drum set. You can record a drum set with three microphones, but eight gives you more control over the sound.

The type of computer needed for music production depends on your production goals. Electronic musicians often prefer to work with laptops, because they can carry their projects to gigs and work on location. The songwriter might prefer a laptop because they can travel to different locations to record instrumentalists and vocalists. They may not have an ideal space for recording acoustic instruments, but perhaps the vocalist has a better location. The film composer will run multiple sample libraries, which require additional storage and computer processing power. They may prefer a desktop because they can add storage to the computer easily as their needs grow. Your current computer may be adequate to meet the needs of everyday computing but lacks the speed and storage required for audio processing. You may end up needing to purchase a new computer.

Your choice of DAW application also depends on the music you want to produce. The electronic musician may choose a DAW application that allows for greater flexibility when working with loops and samples. They may need to import audio files and quickly adjust the tempo to match their current project. When performing, they may need an application that works in the studio and on stage. The songwriter may prefer an application with excellent audio processing and editing. They may also require additional tools to remove background noise. A film composer will spend a lot of time creating MIDI sequences and need to adjust notes to affect their tone and dynamics. The film composer will prefer an application with extensive MIDI features to speed up their workflow.

I hope these examples illustrate how the type of music your produce affects your audio hardware and software needs. If you want to experiment with different types of music, then you can start with a basic configuration and add equipment as your needs evolve. There is no single solution that works for everyone. The development of computer audio has given the home user the flexibility to design their own solutions. The quality of audio hardware and software is excellent, regardless of the price points. You can create music of professional quality regardless of your budget.

Acoustics

Before we begin studying concepts and technology, let us take a moment to explore acoustics. Acoustics is the study of how sound travels in our environment, such as a room or concert hall. Acoustics is how we quantify the way a room responds to sound. Some rooms will sound better than others. Musicians may prefer working in one studio over another because of the way the rooms sound. The room that we work in can and will influence how we feel about our work. When we analyze the acoustics of a room, we look at four components: isolation, balance, separation, and ambience. Let us look at each component closely with respect to a typical room in a house. If you wish, go into the room that you are planning on using for your studio and follow along as we examine each component in that space.

Isolation

Isolation prevents external noises from entering the room whether through the air, ground, or building structure. A room with good isolation prevents you from hearing outside noise. If you can hear cars and people walking by outside the window of your room, then your room has poor isolation. If the windows or walls shake when a large truck passes by on the street, your room has poor isolation. Isolation also prevents sounds within the room from being heard outside of the room. If you intend to record drums in your room and the neighbor across the street can hear your drums, then your room has poor isolation. The reality is that most rooms in a house or apartment will have poor isolation.

This is a reality we will need to accept as we build our home studio. We will have to consider this when we record acoustic instruments so that we do not disturb the neighbors and capture as little outside noise as possible. Headphones help isolate us from outside sounds and keep us from bothering others. Instead of purchasing large speakers that must be placed further away from us for proper balance, we can use smaller speakers that are placed closer to our ears. To make matters worse, there are structural matters we need to consider.

I once lived in a place where the electrical grounding of the sockets was poor. Ungrounded outlets can transmit electrical noise to your recordings. A microphone can pick up the noise as can speakers. Anytime the refrigerator ran, I would hear a hum in my headphones. The electrical noise even traveled along the microphone cable resulting in an electrical hum in my recordings. I did not have the means to ground every outlet in the home, so my practical solution was to unplug the refrigerator any time I wanted to record. It might seem like a silly solution, but it worked.

Balance

Balance refers to the frequency balance in the room. If you play an acoustic guitar in the room, does the guitar sound muffled or bright? Do your speakers sound like they have too much bass or not enough bass? A balanced room does not adversely affect the acoustic balance of instruments or speakers. A room with a bed against a wall and thick curtains will change the frequency balance of instruments and speakers. An empty room with no furniture will also affect frequency balance.

The shape of a room also affects frequency balance. A typical room in a house has 90-degree angles in each corner and sometimes the ceiling. Right angles cause sound to reflect back to the source. These reflections will affect the frequency balance negatively. Professional recording studios avoid right angles and parallel walls. Rooms in a home are squares and rectangles.

There are acoustic solutions in the form of foam panels and wedges that you can place in corners and walls to reduce the number of reflections in a room. However, unless you analyze the room, you are guessing when to place these products. In the end, you could end up doing more harm than good with these products. These products must be taped or glued to the walls and corners. If you are renting the place you live, your landlord may not appreciate your attempts to improve the acoustics of your room.

Separation

Separation is a room's ability to keep different sounds, like instruments, from interfering with each other. If you record two guitars in your room, will the microphone on the first guitar pick up sound from the second guitar? If you have a piano in the room playing with the guitar, can you place the instruments far enough apart so that their sounds do not interfere with each other? Chances are you will not be able to record two separate instruments at once in your space. You could place musicians in different rooms, which is how professional recording studios operate. Studios have isolation booths that allow them to place musicians in their own soundproof rooms. Placing musicians in different rooms in your home will create a degree of separation. You still have the isolation problem, but that variable is unavoidable in a home.

Ambience

Ambience, like balance, affects how an instrument or voice sounds in a room. The ambience creates a sense of space. When you sing in a shower, the tiled floor and walls create an ambience that makes the sound of your voice large and reverberant. The sound of a guitar playing in a room with a flat ceiling will be different than in a room with a large, vaulted ceiling. A room with a bed, carpet, and thick curtains will offer less ambience than a room with minimal furniture and hardwood floors. The ambience of a room should be subtle, but noticeable. A bathroom may have too much ambience but hanging an extra towel in place can reduce the effect. If I need to record vocals in my home, I will often use the bathroom because of the ambience. If the room feels too reflective, I hang towels in different places until I achieve a suitable ambience for the voice.

Your Room

After considering the acoustics of your room, you might feel that creating a home studio might not be a great idea. This analysis is not meant to discourage you, but rather make you aware of your working space. If you are aware of the limitations of your space, you can learn to work with and around those limitations. The type of music you wish to create is also a factor to consider. If your interest is electronic music, and you work with virtual instruments on your computer, then recording acoustic instruments is probably not a high priority. A songwriter may decide to record an acoustic guitar using the pickup on the instrument rather than a microphone. As you work in your room, you will discover ways to adapt to the space to achieve your desired results. Many musicians create music in less-than-ideal spaces; the key is to focus on creativity and not the problems.

Summary

Now that we have a better idea of the room that we will work in along with the equipment we might need, let us proceed with understanding the concepts and equipment involved with music technology. This book is an introduction to music technology and will provide you with basic and practical information to get you started. This book will guide you through essential concepts and technology associated with music production and give you the confidence to make informed decisions about your purchases. You may find yourself asking more questions at the end of the book and that is intentional. I hope to capture your interest and engage your curiosity to learn more about music technology.

At the end of each chapter, you will see an Activity section. The goal of the Activity is to provide you with tools to help you understand the concepts and technology presented in the chapter. Sometimes the Activity will be a series of questions to help you consider your options. Other times, we will install free software to experiment with the material presented in the chapter. All the information I share with you in this book has a purpose, and I want to make sure you are aware of that purpose.

This book will present many new terms for some of you. To help you remember the important terms, each chapter ends with a list of definitions presented in the chapter. This way, if I use a term in Chapter 6 that was explained in Chapter 5, you can easily go to the Terms section and quickly refresh your memory.

Now that you understand the purpose of this book, we can begin our journey.

Activity 1

Introduction

This first activity is to help you determine your music production goals. Having a clear idea of your production goals will help you determine the type of equipment you will need for your home studio. The genre of music you wish to create is one variable. However, the means that you wish to go about creating music in that genre is more important when considering your equipment. I would like to share with you my recommended system. This system will serve most users and is a good place to start.

- Windows or Macintosh desktop or laptop no older than three years
- USB audio interface with two inputs into outputs
- Dynamic or condenser microphone
- MIDI controller with 25–61 keys
- A pair of speakers or headphones

We will look at each of these components in close detail in this book. For now, I am presenting you with this suggested system so that you can begin to look at your investment and how much you will need to spend. Having this as a starting point is helpful when creating a budget. Now let us look at our production goals to see how they influence our choices.

Producers

There are four types of producers that I wish to describe. You will notice that the type of music that these producers create is not mentioned. What is important is the creative process of each of these producers. As you read through these descriptions, think about which one best describes the way you work or want to work.

Self-Contained Producer

The self-contained producer is a solo act. This person handles all the music recording and production. This individual may need the flexibility to work in different environments, especially if they intend to perform their music in front of an audience. This type of producer will tend to work "in the box" meaning that all the sounds come within the computer and the DAW application. There might be occasions when a guest artist is invited to participate in the session. This guest may provide their own tracks and share them with the producer, or they may wish to record live material on existing tracks. Therefore, this type of producer needs some flexibility to add live performances from time to time. This producer may prefer a controller that offers keys for entering melodies and chords as well as pads for entering percussion parts.

Songwriter Producer

The songwriter may perform their music with acoustic instruments, and either be a solo act or work with other musicians. Anyone working with acoustic instruments will prefer having a system that accurately captures their performances on these instruments. This means that they may prefer higher-quality microphones and a selection of microphones for different instruments. This type of producer may prefer to work alone during the creative process and then add musicians later. This producer may need virtual instruments to create temporary tracks for the musicians to follow. This individual may also want to perform songs live using backing tracks on a computer. If the songwriter is a pianist, then they may prefer a controller with 88 keys and something with the feel of the piano.

Composer Producer

This individual creates music for media such as film, television, and games. The composer may be asked to write in a variety of genres and therefore needs access to a large collection of sounds. Virtual instruments such as orchestras and pianos are important to the composer because they may be asked to create a mock or temporary score for a media production. This person may also be asked to create notated parts for musicians to create the final version of the score. They may also need a collection of synthesizers for other productions. When working with film, their computer must be powerful enough to play back video and remain synchronized to the music they are creating. The composer will often prefer working with a full-size controller with 88 keys so that they can cover the range of an orchestra easily.

Performer Producer

The performer producer is more interested in live performance. This individual sees the studio as a sketchpad for ideas to collaborate. The performer needs to accommodate multiple musicians at once and provide everyone with a comfortable listening environment. This individual needs a powerful computer to capture multiple tracks at the same time. Depending on the type of music being created, an assorted collection of virtual instruments is required to work in a variety of genres. The performer may find themselves doing solo performances, in which case they need a portable system that can travel with them. The type of controller that this person uses depends on the type of music they wish to perform.

Assessment

To help you determine your production style, answer the following questions.

How do you prefer to work on your productions?

- Alone
- With others
- Combination of both

Do your productions require live instruments?

- No
- Often
- Sometimes

What type of sounds do you use for your productions?

- Electronic
- Acoustic
- Both

What is your skill level with playing instruments?

- Minimal
- Average
- Extensive

What type of work environment do you like to work in?

- At home in a single location
- Multiple locations
- Combination

What is your experience with producing music?

- Minimal
- Some
- Extensive

Conclusion

The goal of this activity is to get you thinking about your music productions. If you are not certain about your answers, that is okay. We are not looking for concrete solutions right now. Instead, we are exploring our options and thinking about how we produce and work on music. Please keep this in mind as we go through this book. As we explore different technologies required for music production, reflect on your own production goals when determining the type and how much equipment you need.

Terms

Acoustics The scientific study of how sound interacts with spaces such as rooms, concert halls, and arenas.

Ambience The amount of sound that persists in a room or space once the source stops producing sound.

Audio Interface A computer device that allows audio signals in and out of the computer for recording and playback.

Balance The accuracy that a room or space represents all frequencies from instruments or voices.

Digital Audio Workstation (DAW) A computer-based system that allows for digital recording and playback of audio. The system consists of a computer, audio interface, and audio application.

Isolation The degree to which a space is isolated from external sounds.

Microphone An electronic device that converts acoustic sound into an electronic signal so that it can be recorded by an analog or digital system.

MIDI Controller A computer device that allows MIDI data to be sent to the computer to control MIDI devices.

Mixing Console An electronic device designed to combine audio signals for recording and mixing.

Musical Instrument Digital Interface (MIDI) A protocol that allows equipped devices to share note, patch, and other programming data for music recording and playback.

Producer An individual who creates or assembles musical elements to create a music composition or song.

Sampler A music device that uses recordings of acoustic instruments to recreate the instruments. A sampler can also use recordings of other sounds.

Separation The degree at which each sound source in a room or space can be clearly heard.

Synthesizer A music device that produces sounds by combining electronic frequencies through a variety of methods.

Chapter 2

What Is Sound

Sound Properties

Before we explore the technology for our home studio, we need an understanding of sound. Everything we do involves sound. We need to understand how sound works, how to measure sound, and how it interacts with our environment and affects our ears. All the terms and concepts we cover in this chapter will appear in nearly every other chapter of the book. Therefore, it is important that we clearly understand the properties of sound.

Air Molecules

For sound to exist, we need a medium; something that sound can exist within. Air molecules are the medium for sound. Air molecules are in constant motion around us and produce pressure in all directions. We may not always be aware of the pressure of air molecules, but if you have ever been in an airplane or traveled to high-altitude areas, you feel the pressure in your ears. This is caused by a shift in air pressure between your inner ear and the outer ear. Like all objects, air molecules are affected by factors like temperature and altitude. Although air molecules are constantly moving, they are all attempting to return to their original state or location. If we look at a lake, we notice that there is constant motion in the water made up of small waves and eddies. If you throw a large stone into the lake, the motion of the water changes, but after a few moments, the motion returns to the way it was. Air molecules behave in the same way.

Imagine holding a blown-up balloon in your hands. The air molecules are in a state of rest. The air pressure in the balloon is equally distributed. Now squeeze the balloon with both hands. The balloon changes shape due to the pressure in your hands. Notice the pressure of the balloon pushing out against your hands as it fights back to retain its original shape. The pressure from your hands is pressing the air molecules together and the molecules are struggling to return to their original state. When you release the pressure on the balloon, it immediately returns to its original shape.

DOI: 10.4324/9781003345138-3

The air molecules inside the balloon experience two different states: compression and rarefaction. Compression occurs when air molecules are pushed together, creating an area of high pressure. This occurs when you squeeze the balloon on both sides with your hands. You feel the pressure increasing the harder you squeeze as the air molecules struggle to return to their original state. Rarefaction occurs when you let go of the balloon. The pressure of the air molecules being pressed together is released and the molecules move away from each other. Rarefaction occurs when air molecules push away from each other, creating low pressure. As the molecules push away, the shape of the balloon restores. The air molecules return to their original state with equal pressure in all directions. The air molecules around us behave the same, although in a much larger space. This constant change between compression and rarefaction is called harmonic motion.

Even though the air around us is in constant harmonic motion, we cannot hear the sound until the sound is created, transmitted, and received. Three components are essential for sound to exist. The first is generation, which is how sound is created. The second is propagation, which is how sound is transmitted. The third is reception, which is how sound is received. These three components are self-explanatory but let us explore them a little closer.

When someone speaks to you, their vocal cords move the air molecules near their mouth, causing them to compress. Vocal cords generate sound by moving the air molecules. The air molecules run into each other and with each collision, the motion propagates to the next one; the first molecule pushes the second one which, in turn, pushes the third one; and so forth. This motion will continue until the air molecules hit a surface like a wall or your eardrum. Your ear canal receives the air molecules and channels them to your eardrum. Your eardrum vibrates back and forth, creating an electrical impulse that is transmitted to your brain. Your brain recognizes the information and the sound of someone speaking. Your ears are responsible for the reception of sound. If I place a microphone in front of the person's mouth, then the microphone becomes the receptor, which converts the sound into electrical impulses. Those impulses enter your computer via an audio interface, which digitizes the sound for storage on a computer.

Any sound source that displaces air molecules is a sound generator. This ranges from the human voice to musical instruments, and even to devices such as jack hammers and jet engines. Listen to the environment around you. The noises you hear from engines, birds, and rustling leaves are all caused by air displaced air molecules. We spend most of our lives learning how to differentiate sounds from one another.

Sound Waves

At this stage, you may be asking about sound waves. We talk about sound moving through the air and experiencing compression and rarefaction, but how exactly does the sound move? Do the air molecules travel up and down, back and forth, or side to side?

Sound energy travels in the form of sound waves. There are two types of waves in nature: transverse and longitudinal. Transverse waves move in an up-and-down motion. If two people hold a rope between them and one person raises and lowers their arm quickly, the rope will create transverse waves (Figure 2.1).

Longitudinal waves travel in a parallel motion, or back and forth. Imagine sitting in a pool of water and using your hands to push the water forward to create a longitudinal wave. You will notice that the waveform keeps pushing against the water and moves along the surface of the water until it encounters a reflective surface such as the wall. If there is enough force when the water strikes the wall, the wave bounces back and returns toward you with less energy. When you speak, the air molecules near your mouth are pushed forward. Those molecules then push the ones in front of them forward. When the molecules reach a reflective surface, like a wall in the room, the sound waves will travel back toward their origin. Sound energy travels as longitudinal waves (Figure 2.2).

However, if you have ever looked at a sound wave as it is represented by audio software on your computer, you might see the wave represented by an image like the one given in Figure 2.3.

The image might lead us to believe that sound waves are transverse even though I just stated that they are longitudinal. The main reason sound waves

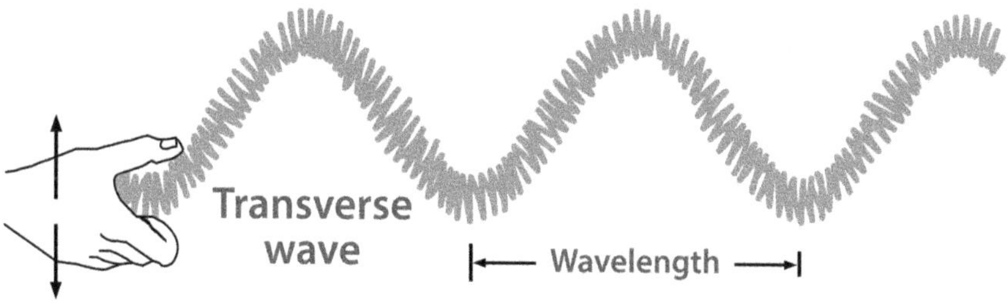

Figure 2.1 Transverse waveform

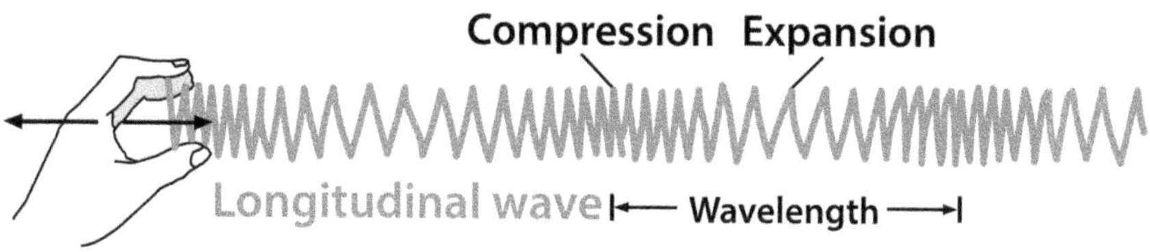

Figure 2.2 Longitudinal waveform

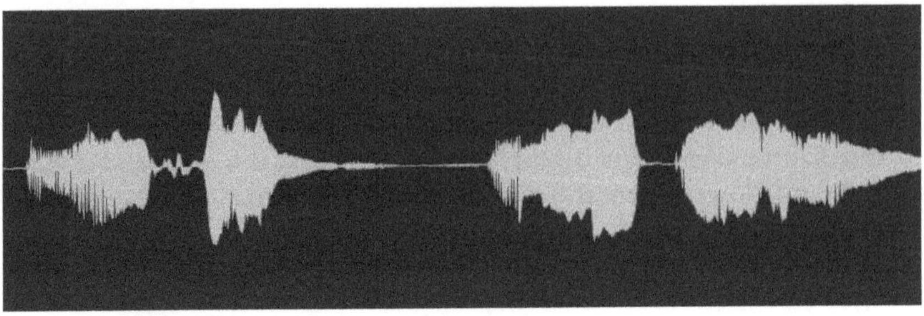

Figure 2.3 Waveform as displayed in audio software

are represented as we see in the image is because of practicality; it is easier to see sound waves represented as transverse. A transverse wave is large when a sound is loud and smaller when the sound is quieter. Sound waves represented as transverse waves are also easier to edit and locate specific points. While this is the way that software represents sound as transverse waveforms, it is not how sound moves through the air. Keeping things practical, we will continue to represent sound waves as transverse waveforms.

Waveforms

Now that we understand sound waves, what they are, and how they are represented, let us look at their specific properties. Waveforms carry two types of information: pitch and amplitude. Pitch describes how highness or lowness of a tone. A child's voice will have a high pitch whereas an adult male will have a low pitch. The right edge of a piano produces high pitches while the left edge produces low pitches. Musicians assign note names to pitches. Audio engineers assign frequencies to pitches.

Pitch

Sound waves vibrate in the air in cycles, and we measure the vibration in hertz (Hz). The unit measurement of hertz is named after scientist, Heinrich Hertz, who was able to prove the existence of cycles in waveforms. A waveform that produces one vibration cycle in a second is described as having a frequency of 1 Hz. The cycle is represented by a line starting at a center point, rising above the point, crossing below the point, and then stopping at the center (Figure 2.4).

A waveform with a frequency of 60 Hz cycles 60 times in one second. A waveform with a frequency of 1,000 Hz (1 kHz) cycles 1,000 times in a second. The numbers seem arbitrary, and it is hard to attach meaning to them. Fortunately, we can use the musical pitch as a reference against hertz and

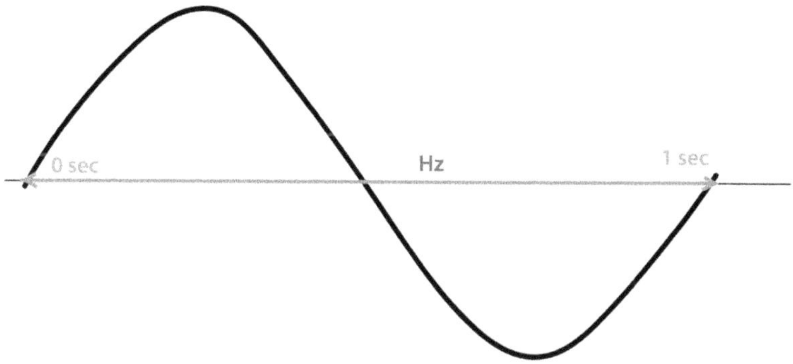

Figure 2.4 Single-cycle waveform

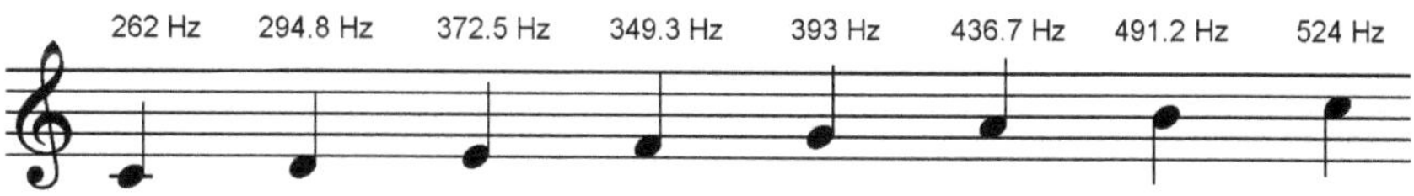

Figure 2.5 Notated C major scale and the related frequencies for each pitch

start to put these numbers into perspective. Musicians will often tune their instruments to the pitch A above middle C on the piano. Musicians will often refer to this pitch as A 440. The number 440 is the number of cycles per second needed to produce the pitch A. Therefore, 440 Hz is the same as A above middle C. It would be convenient if every pitch related directly to hertz with a whole number, but this is not the case. The diagram below shows a C major scale and the number of hertz each note produces. The only whole numbers in the scale are G and A (Figure 2.5).

Bandwidth

Some of you might be curious as to what are the lowest and highest frequencies we can hear. In other words, we would like to know the frequency range, or bandwidth, of our hearing. The bandwidth of human hearing is from 20 Hz to 20,000 Hz (20 kHz). Frequencies above 20 kHz are called ultrasonic while frequencies below 20 Hz are called infrasonic. Bandwidth can also describe the range of frequencies that an instrument or voice can produce. A piano has a bandwidth of 27 Hz–4,100 Hz, while a guitar ranges from 82 Hz to 2,100 Hz. Table 2.1 shows the approximate bandwidth of other musical instruments.

Table 2.1 Frequency range of instruments

Guitar	*82 Hz–2,100 Hz*
Piano	27 Hz–4,100 Hz
Tuba	41 Hz–440 Hz
Flute	261 Hz–1,975 Hz
Bass guitar	31 Hz–450 Hz
Human voice	87 Hz–900 Hz
Violin	196 Hz–4,500 Hz

Amplitude

When we describe how loud a sound is, we are referring to its amplitude. The amplitude of a waveform determines how many air molecules are displaced by the sound. A loud sound will displace a lot of air molecules while a quieter sound will displace far less. The more air molecules displaced, the greater the sound pressure level. If we go back to our example of squeezing a balloon, applying more force compresses the air molecules more and you feel greater pressure against your hands. Applying less force displaces fewer air molecules creating less pressure. Going back to our waveform illustration, the amplitude of a sound is represented by the height of the waveform. A waveform diagram that goes far above and below the center line has a high amplitude (Figure 2.6).

We measure the amplitude of sound in decibels (dB). The term is named after Alexander Graham Bell, who along with developing the telephone, was trying to find a way to measure the loss of amplitude as sound traveled along a phone line over great distances. Measuring amplitude is tricky because a decibel is not always an absolute value; sometimes it is a reference point. Frequency is absolute because a 10 Hz waveform is always the same. A sound that is 100 dB in amplitude varies depending on whether we are measuring the sound pressure level in air molecules (dB SPL) or the voltage produced by an audio system (dBv). To complicate matters, digital systems, such as your computer use a digital scale to measure amplitude (dBFS). The biggest difference between these three is that 0 dB means something different for each of them. For now, we will focus on dB SPL because it is the measurement that we can relate to the easiest and therefore will mean the most to us.

When we determine the amplitude of the sounds we hear naturally in our environment, we can measure the sound pressure level (SPL), which describes the pressure of the air molecules displaced. An SPL meter is a device that has a microphone to capture the sound in the environment and then display the amplitude of the sound in dB SPL. You can purchase an SPL app for your mobile phone as well. It is not as accurate as a dedicated device, but it is enough to provide you with a useful reference when entering loud environments.

Figure 2.6 Loud waveform compared to a quiet waveform

Since we can define the range of hearing in frequency we might as well define the range of hearing in decibels. The range from the quietest to the loudest sound that a person, instrument, or device can produce, or capture is called dynamic range. Dynamic range is measured in decibels. The dynamic range of our hearing starts at 0 dB, which is called the threshold of hearing, and ends at 120 dB, known as the threshold of pain. We define the threshold of pain as the point where most people will experience physical pain when a sound is too loud. Table 2.2 shows the amplitude of common items in dB SPL.

The table tells us a quiet room has an amplitude of 40 dB, while a normal conversation is roughly around 60 dB. If you are standing at the corner of a busy intersection, then you might experience sound around 75 dB. While a hair dryer is 85 dB. How loud is 85 dB? The rule of thumb is that if you are three feet away from someone and you must shout for them to hear you, then the environment that you are in is around 85 dB. Keep that number in mind when we talk about your ears and hearing loss later in this chapter.

Speed

We know that sound is created when air molecules are displaced, but we have yet to address the speed at which sound travels. You might have heard someone say that a jet plane just broke the sound barrier, which can cause you to imagine a wall of sound that a jet smashed through. Well, there is

Table 2.2 Loudness of common items in dB SPL

Quiet room	*40 dB*
Normal conversation	60 dB
Traffic noise	75 dB
Vacuum cleaner	80 dB
Hair dryer	85 dB
Loud concert or club	100 dB
Jet engine	120 dB
Jackhammer	130 dB
Gunshot or fireworks	140 dB

some validity to that idea. At a median temperature of 70 degrees Fahrenheit, sound travels at 1,130 feet per second or 770 miles per hour. Breaking the sound barrier means traveling faster than the speed of sound. Breaking the sound barrier creates a sonic boom, which is a loud explosion of sound. Remember that air molecules are all around us and moving back and forth at different speeds. When those molecules are compressed and produce sound, they travel at 770 miles per hour, colliding with other molecules until the sound reaches your ear. The air molecules are dense and essentially create a physical barrier. As a plane approaches 770 miles per hour, the air molecules are compressed quickly and become very dense. Once the plane crosses 770 miles per hour and moves through this dense collection of air molecules, the pressure quickly dissipates (rarefaction) resulting in a loud thunderclap-like sound, like an explosion. The plane is literally smashing through a wall of air molecules.

Wavelength

If we take our single-cycle waveform and stretch it out so it is a straight line, then we can measure the length of the waveform. The length of a waveform (W) is determined by the speed of sound in feet per second (v), divided by the frequency of the waveform (f). This can be represented by the formula W = v/f. Using this formula, we determine that a 1,000 Hz waveform is equal to 1.13 feet (W = 1,130/1,000). A 2,000 Hz waveform has a length of 0.556 feet or half the length of a 1,000 Hz sound. A 500 Hz waveform has a length of 2.26 feet, twice the length of the 1,000 Hz waveform. Therefore, we can conclude that higher frequencies have shorter wavelengths while lower frequencies have longer wavelengths.

Wavelengths can help us determine how much energy is required to produce a frequency. If we imagine these waveforms as lengths of rope, think about how much force is required for you to pick up the rope and move it up and down to create a waveform. The shorter rope would be much easier than the longer rope. A 2,000 Hz waveform is a little over half a foot long while a 200 Hz sound is around five and a half feet in length. A 55 Hz waveform is nearly 20.52 feet in length. It takes more energy to produce a 55 Hz waveform than one at 2,000 Hz. Low frequencies require a powerful amplifier to produce sound. High frequencies can be produced with a much smaller amplifier. Keep this in mind when we talk about loudspeakers in a later chapter.

Waveform Types

Complex Waveforms

We can separate waveforms into two categories: complex and simple. Complex waveforms are sounds that occur naturally or acoustically. This can be the sound of birds outside or the sound of a gasoline engine accelerating. They can be the sound of an acoustic piano, and electric guitar, or the human voice. Complex waveforms are not repetitive; they are constantly changing. You may play the same note on a piano, but each time you press the key, the pressure in your finger is slightly different, resulting in a different waveform. It still sounds like a piano and the pitch is the same, but there are subtle differences in each repetition. The subtleties are what make the waveform complex.

When you play A 440 on a piano, you are hearing more than just the piano playing a pitch at 440 Hz. There are additional frequencies that we perceive that are not part of the original sound. These additional frequencies occur above the initial sound, called the fundamental. These additional frequencies, or harmonics, add color and dimension to the sound. The harmonics, combined with the fundamental, create the timbre of the sound. Timbre is what allows us to distinguish the sound of a piano from that of a guitar. The harmonics in a piano are created by the soundboard, the lid, the neighboring strings, and the felt hammer striking the string; all the elements that make a piano are responsible for the harmonics.

These harmonics occur naturally. In fact, they are mathematically predictable. If I want to know the first harmonic above A 440, I multiply it by two and end up with 880 Hz. The fourth harmonic above A 440 is 440 times 5, or 2,200 Hz. Harmonics were discovered by the Greek philosopher Pythagoras during the 6th century BC. He made the discovery using a monochord, a plucked instrument akin to a harp but with a single string. You can experiment discovering harmonics on your own using an acoustic guitar. The lowest open string on a guitar is an E. Pluck the string once and look at the other five strings. You will see that they each are vibrating. The vibration of these strings creates subtle, yet crucial harmonics. Now place your finger over the 12th fret of the instrument, which is half the length of the string on most guitars. If you lightly place your finger above the fret and pluck the string, you will hear the note E, but an octave higher, or twice the frequency. If you place your finger at the quarter length of the string, over the fifth fret, you will hear a note four times the frequency of the low E. You can create harmonics on nearly every fret on the string. Some of the harmonics will sound in tune, while others will not. When you pluck the low string on its own, each of these harmonics is produced at different volumes, all contributing to the timbre of the guitar.

Each instrument produces its own set of harmonics, which give instruments their color and personality. The materials that are used to create instruments help shape the sound that it produces. An acoustic guitar with a rosewood body will sound different from one with a spruce body. They both sound like guitars, but these different woods produce harmonics at different volumes, which alters the timbre. Timbre is what makes sounds interesting to our ears.

Simple Waveforms

Simple waveforms are waveforms that repeat in predictable ways. The waveform is identical in every instance. These waveforms are sometimes called periodic, which means that occurring at regular intervals. Simple waveforms are created electronically and are often used to test electronic equipment because the waveforms are consistent and predictable.

There are three basic simple waveforms that we need to be familiar with. Each of these waveforms has a different shape, producing different harmonics, which helps us distinguish them aurally. The first and simplest is the sine wave. The sine wave represents the purest tone possible because it has no harmonics. When you hear a sine wave, you are hearing the fundamental and nothing else. The sine wave is the simplest representation of a waveform, which we have seen several times in this chapter (Figure 2.7).

The sawtooth wave is the most complex of the simple waveforms, which almost sounds like a contradiction. The timbre of a sawtooth wave is defined by the sound of fundamental along with odd and even harmonics. The first harmonic is about half the volume of the fundamental. The additional harmonics appear at half the volume of the previous one. The sawtooth

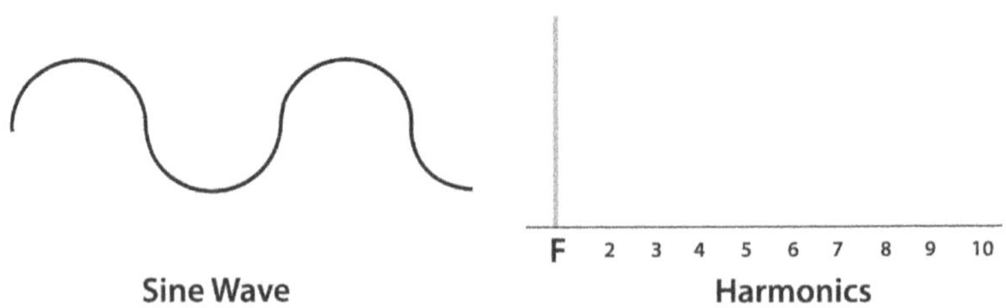

Figure 2.7 Two cycles of a sine wave with harmonics

waveform has a distinct buzzing tone that is popular with sound designers because of the timbre, which can be shaped with filters. The shape of the waveform resembles the teeth of a sawtooth blade (Figure 2.8).

The square wave also contains harmonics, but at a lesser degree than the sawtooth. Only the odd harmonics are present in a square wave, each at half the volume than the previous one. The resulting timber is less dramatic than the sawtooth and almost hollow sounding. The square wave is similar in character to the sawtooth wave (Figure 2.9).

Noise

Along with simple waveforms, noise is used to test and measure electronic equipment. There are two types of noise used for testing audio equipment such as microphones, loudspeakers, and headphones.

The first type is called white noise. White noise consists of mixed and random frequencies. All the frequencies sound with the same intensity. The resulting sound is aggressive and irritating. Using white noise requires care as you can damage electronic equipment. You will want to turn down your speaker volume before listening to the example on the companion website.

Pink noise is subtler than white noise but still intense. Pink noise also consists of all frequencies, but the intensity of the frequencies tapers as the

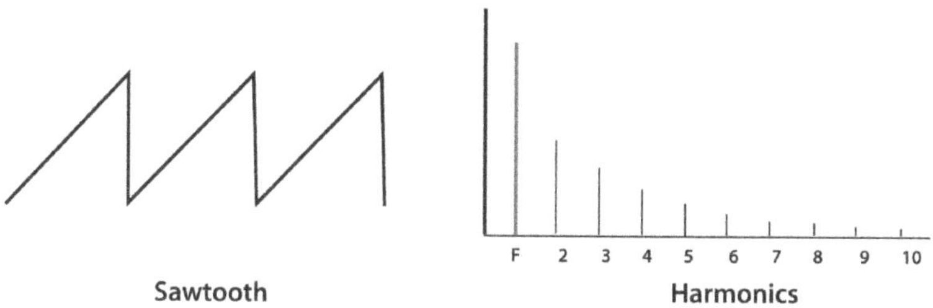

Figure 2.8 Three cycles of a sawtooth wave with harmonics

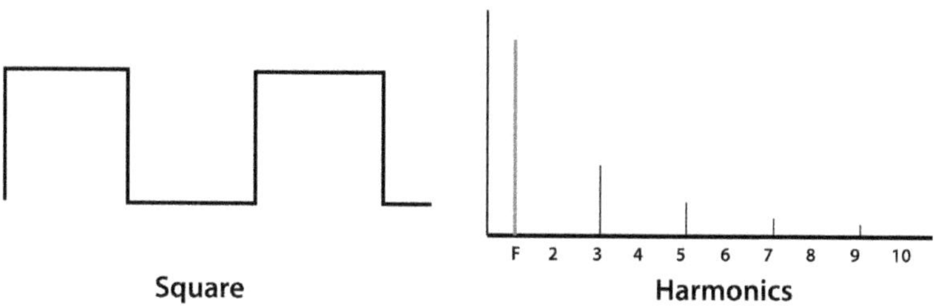

Figure 2.9 Two cycles of a square wave with harmonics

frequencies rise; only the octaves share the same intensity. This means higher frequencies are not as loud as the lower ones. The result is a less aggressive sound, which can be filtered to imitate the sound of the ocean or wind. Noise machines used to help one sleep use filtered pink noise to create soothing and regular sounds to help one relax.

In the Activities section of this chapter, we will use a free audio editor program called Audacity to create simple waveforms along with pink and white noise so that we can experience their tones.

Phase

Every image of a single-cycle waveform shown in this chapter always has the waveform starting at the center line and rising up followed by the same shape occurring underneath the center line before returning to the center. If we think of the waveform cycle as a circle that has been cut in half, then we can assign each high and low point of the waveform in degrees of a circle as shown in Figure 2.10.

A waveform that starts at 0 degrees is starting in phase. If the waveform starts at any other degree, then it is starting out of phase. If the waveform starts a quarter of the way in the cycle, then we say that the waveform starts at 90 degrees. A single waveform starting at any degree will not affect the resulting sound or how we perceive it. Phase only matters when there is more than one waveform sounding at the same time.

Phase is the time relationship between two waveforms. If two waveforms begin at the same time, and start from the same point, such as 0°, then the two waveforms are in phase. Two waveforms that are in phase will reinforce each

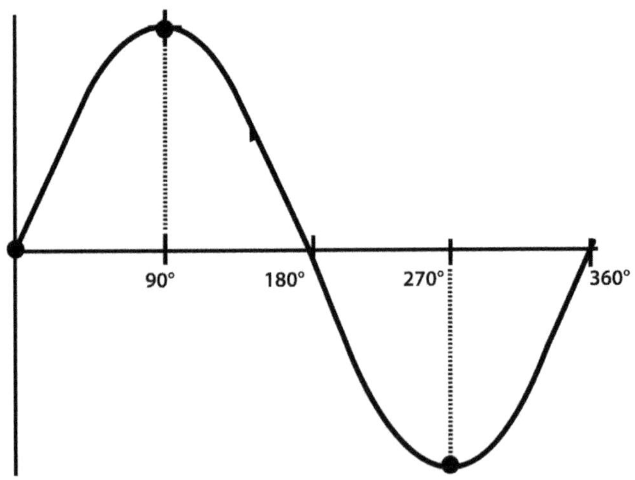

Figure 2.10 Single cycle of a sine wave with degree indications

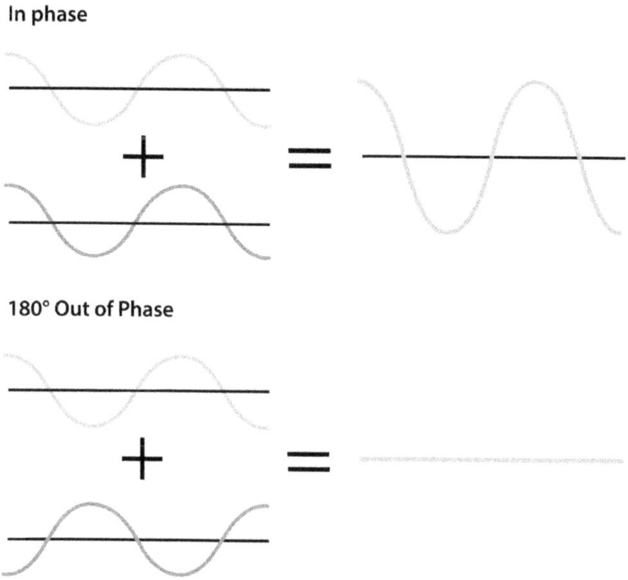

Figure 2.11 Two waveforms in phase and then 180 degrees out of phase

other and increase the overall amplitude of that sound. If two waveforms start at different points, then those waveforms are out of phase. If the first waveform starts at 0 degrees and the second waveform starts at 90 degrees, then the waveforms are 90 degrees out of phase. The resulting amplitude will be weaker than if a single waveform was sounding. If the second waveform is 180 degrees out of phase, then the two waveforms are opposite of each other and therefore will cancel out all frequencies. If both waveforms are sine waves, then we will hear no sound at all (Figure 2.11).

Phase matters when there is more than one waveform sounding at the same time. It also matters when using more than one microphone on a single instrument, such as a piano or guitar. If the microphones are not placed correctly, the resulting sound may experience phase cancellation. We will come back to phase cancellation when we talk about microphone placement in Chapter 7.

Envelope

Along with timbre, the other element that allows us to distinguish one instrument from another is articulation of the sound. Articulation refers to how the amplitude of the sound changes over time. For example, does the sound stay at the same level the entire note is playing? Does the full sound occur at the beginning, or does it gradually come in? What happens after the instrument stops playing, does the sound immediately stop or does it ring out for a little longer? We describe the articulation of sound by looking at the envelope of the waveform. The envelope of a sound describes the time it takes for the amplitude of the sound to change and the amplitude level during or after the change. The envelope of a sound consists of four stages: attack, decay, sustain, and release. We often use the acronym ADSR when explaining the envelope of a sound (Figure 2.12).

The attack variable of the envelope measures the time it takes for the sound to begin. A piano has a fast attack; once you strike the key you immediately hear the sound. A violin has a slow attack because the movement of the bow across the strings takes a moment before there is enough energy to cause the strings to produce sound. A snare drum has a fast attack since sound is produced immediately when the stick strikes the drumhead. A trumpet has a slow attack because enough air must enter the instrument before the sound is produced.

The decay of the envelope measures the time it takes for the amplitude of sound to reach its sustain level while you are still playing the sound. When you play a key on a piano, the initial sound is loud, but the amplitude drops to a lower level as you continue to hold the key down. A violinist has more control over the decay of a sound. The violinist can keep the pressure of the bow the same after the note starts, or they can reduce the pressure and lower the sustain level. A snare drum, however, has no sustain because after you strike the drumhead, you have finished playing the instrument.

The sustain variable of the envelope is the only variable that does not measure time, but rather amplitude level. Not all instruments have a sustain variable in their envelope. Sustain can only exist while the instrument is being played. In the case of the snare drum, you strike the drumhead with a stick,

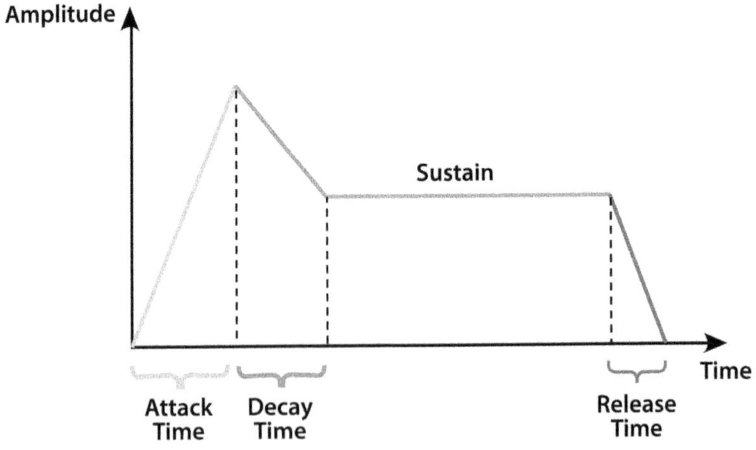

Figure 2.12 Four-stage envelope of sound

and you are done. You can strike the head again, but now you are starting a new envelope. The absence of a sustain variable means that the decay variable is also not present on a snare drum. The sustain level of a piano changes over time. If you hold the key down after striking a note, the string is free to vibrate, but it loses energy over time and will eventually stop sounding. The violin can change the sustain level of a note but vary the pressure of the bow. The trumpet player can sustain the note if they have the breath to do so.

The release measures the time it takes for the sound to fade away after the sound is released. A snare drum will have a fast release unless the drum is resonant and rings out for a long time. The body of a violin will allow the sound to persist after the violinist releases the bow from the string. A trumpet only produces sound when air travels along the tube, so the release time is short for this instrument. The piano has a resonating body like a violin, so the release time will be longer for this instrument.

Spend a little time listening to different instruments and examining the envelope for each. The envelope, along with timbre, helps you discern one instrument from another.

Hearing

Your ears are the most critical component when working with audio. Your ears determine the quality of sound you are producing and recording, how sounds interact with each other, and the frequency balance within the sound. Your ears allow you to distinguish between sounds and determine which one works best in your projects. Your hearing allows you to make critical decisions about sound and shape the result of your work. Without accurate hearing, working with audio can be difficult and frustrating. Therefore, it is important that we protect our hearing and understand how the ear accomplishes its tasks. The human ear consists of three major sections, the outer ear, the middle air, and the inner ear. Each of these sections works together to capture and transmit the sounds around us to our brains.

The outer ear is made up of two parts: the pinna and the auditory canal. The pinna is your earlobe and allows you to determine the direction of sound. We are capable of hearing sound 360 degrees around our heads. Both ears work together to help us determine the direction of sound. The pinna also deflects loud sounds so that they are less damaging to the middle ear. The pinna channels sound to the auditory canal, which funnels the sound to your middle ear. The auditory canal is most sensitive between the frequencies of 2,000 Hz and 5,000 Hz, which is where human speech happens. Our sensitivity to these frequencies allows us to distinguish consonants such as the letters B and P as well as vowels like O and U.

The middle ear is where sound, or acoustic energy, is converted to mechanical energy and then sent to our brain. The tympanic membrane, or eardrum, vibrates back and forth as sound waves strike it. The eardrum is an extremely thin and sensitive membrane that can capture a wide dynamic range. The sensitivity of the membrane means that it can capture extremely quiet sounds, but it can also be permanently damaged by extremely loud sounds. Without any protection other than the earlobe, the eardrum can be easily damaged by a single sound. If the eardrum is constantly stimulated by loud sounds, it loses sensitivity over time, making it difficult to hear quieter sounds. The membrane of the eardrum can be ruptured by loud sounds. The eardrum can heal over time, but the ruptured area is replaced by scar tissue, which negatively affects sensitivity to quieter sounds.

The hammer, anvil, and stirrup work together in the inner ear to capture the movement of the eardrum and transmit that information to the cochlea. The hammer, anvil, and stirrup amplify the movement of the eardrum so that the information sent to the cochlea is clear and strong. The cochlea is coiled, filled with fluid, and has thousands of tiny hairs called the basilar membrane running inside. These tiny hairs respond to the signal transmitted by hammer, anvil, and stirrup and convert the acoustic energy to electrical or neural energy. The information is sent to the brain, where the sound is processed. Sounds we have heard before are immediately recognized whereas new sounds are studied and cataloged for future use.

The ear is an important part of our lives, so we need to examine ways of protecting our ears so that we can always rely on them for our work with audio.

Hearing Loss

We know that the ear is extremely sensitive and can capture frequencies between 20 Hz and 20 kHz. However, the upper range of our hearing range diminishes as we grow older and are exposed to loud environments. The upper limit can drop down to 16,000 Hz by the age of 20 years and even down to 8,000 Hz by the age of 65 years. Our hearing is sensitive enough to detect extremely quiet sounds and the threshold of pain is 120 dB. Once again, these numbers change as we grow older and encounter loud environments. Our hearing is delicate and can be easily damaged if we are not careful.

Earlier in this chapter I showed you a chart with decibel levels are various sounds and even provided you with a quick way of determining when the ambient sound around you is 85 dB SPL or higher. What we need now are some guidelines to help us determine when you need to leave the loud environment or find hearing protection.

In the United States, the Occupational Safety and Health Administration (OSHA) determines safe working conditions for individuals. The office regulates the safety of work environments including hearing. The chart given in Table 2.3 shows how many hours a person can be exposed to different decibel levels before needing hearing protection.

Table 2.3 Allowed daily exposure to loudness

Loudness	*Allowed Daily Exposure*
82 dB	16 hours
85 dB	8 hours
88 dB	4 hours
91 dB	2 hours
94 dB	1 hour
97 dB	30 minutes
100 dB	15 minutes
103 dB	7.5 minutes

Following these guidelines, if you are with a group of people in a loud restaurant and you are all shouting to hear each then the ambient noise is 85 dB or higher. OSHA regulations recommend that you should find hearing protection after eight hours. Keep in mind that OSHA regulations are designed for the general population. If you plan to work in audio, you should probably be stricter than OSHA regulations. This means, if you know that you are going to a loud restaurant or club, you should wear hearing protection as soon as you enter the space. There is no need to wait several hours before using hearing protection.

Let me share a personal story that you may have experienced as well. You and your friends attend a live concert in an arena. You are aware that the sound level is over 85 dB because you are all shouting at each other throughout the two-hour event. After the concert, you drive home listening to the radio. You arrive home and go to bed, noticing that your ears are slightly ringing. The next morning you wake up and the ringing in your ear is gone. However, when you enter your car, you find that the volume level of the radio is extremely high and almost painful. What happened here?

The night before, you experienced what is called a temporary threshold shift. This is a reduction in sensitivity to sound levels. The longer you expose your hearing to loud sound, the less sensitive it becomes. Subconsciously, your brain adjusts your hearing so that the loud environment is the new normal level of sound. Your hearing becomes less sensitive and quiet sounds feel quieter. Your normal radio volume seemed quieter on the drive home, so you turned up the volume. Your ear readjusts overnight and returns to its regular sensitivity in the morning. This is the reason why your car radio suddenly feels extremely loud. The shift was temporary, and thankfully, your hearing returned to normal.

However, if you continue subjecting your ears to this kind of environment, one day you may wake up and the ringing in your ears is still present. You may also notice that everything around you is quieter, and you need people to speak louder to you in order for them to hear you. This is known as a permanent threshold shift, or hearing loss.

Another type of hearing damage that can occur is called tinnitus. This condition is the continuous perception of noise or ringing in the ear even when there is no sound. The sound can be a hum, ringing, or another noise that is constantly present. The constant sound is most noticeable when you are in quiet environments. Tinnitus cannot be cured. Hearing loss cannot be cured. Your brain is dynamic and will adjust to your hearing situation, but you may find it more difficult to work with audio because you will not be able to accurately hear frequencies and dynamics. Fortunately, there are some cost-effective ways to prevent all of this from happening.

You should make every effort to avoid exposure to loud environments for long periods of time. If you must be in a loud environment, then you should wear earplugs. There are several companies selling earplugs that are designed for musicians and promise a natural sound. Some of these solutions are affordable, others are expensive. These devices might be ideal if you are a live mixing engineer working in arenas and stadiums. If you are just an audience member at a concert you might find it very difficult to justify the expense of these types of hearing protection.

Consumer foam earplugs are designed to help people sleep when they live in loud environments. These earplugs provide sufficient protection in loud environments. You can purchase these earplugs in bulk for a low cost. When you first wear foam earplugs, the sounds you hear might be muffled and hard to distinguish. However, your ears are sensitive, and your brain will quickly adjust to the environment. Within a short period of time, you will find that you are able to hear almost as clearly with the earplugs inserted as you could without them. You can wear these earplugs at a loud event and still enjoy the event and drive home without any ringing in your ears. The other advantage of buying these earplugs in bulk is that you can keep a pair everywhere you go, and they are easy to replace if lost.

You should also avoid using any type of earbuds or in-ear headphones to listen to music. Remember that the outer ear is designed to help filter and deflect sound so that it is not striking your eardrum at full force. Using earbuds bypasses the outer ear and allows you to send sound directly to the eardrum. Since earbuds do not fully isolate you from external sounds, you will probably increase the output volume so you can hear the music clearly. The three-foot rule still applies in this situation. If you are three feet away from someone and they can hear the music from your earbuds, then you are causing hearing damage. If you are wearing earbuds and cannot hear someone talking to you from three feet away, then you are causing hearing damage. Earbuds are extremely convenient, lightweight, and less bulky than standard headphones. They are ideal for phone conversations but not much more. If you must use earbuds to listen to music, then make sure to keep the volume low. Hearing damage is permanent and not reversible.

Summary

Understanding sound and hearing is critical if we intend to work in the audio field. In this chapter, I have given you the details you need to describe the properties and components of sound. You understand how to measure the frequency and amplitude of sound. You are aware that sound has harmonics that help shape and determine its timbre. You can explain the difference between complex and simple waveforms. You can describe the envelope of sound and how instruments differ in their envelopes. You are aware of the parts of the ear and their functions. Finally, you know how to protect your ears so that you can continue to create and produce music for as long as you wish. Now that we have a foundation in sound, we can apply this knowledge to the information that follows in the remaining chapters.

Activity 2

Creating Waveforms

In this activity, we will use a free audio editing application, Audacity, to create different waveforms. You will be able to see and hear each waveform type Audacity. Audacity is a free audio application that runs on Windows and Apple computers. The system requirements for Audacity are minimal, so it can run on older computers. Audacity does not require specific hardware for audio; as long as your computer has a sound card, Audacity will play back audio.

Please visit https://www.audacityteam.org/ to download Audacity. Clicking on the Download Audacity button will load a new page. The web page will detect your operating system and automatically download the correct version for your computer. The file will download to your computer's default location for file downloads, which is usually a folder called Downloads in your home folder.

Locate the installer file and double-click on it to launch the installer. Follow the prompts on your screen to install the application. Once the application is installed, launch the application. Audacity will automatically detect your sound card and configure it to run on your computer. Once the program starts, you can begin the Activity.

After you complete the Activity, feel free to experiment with Audacity to see what other audio functions it can perform. It is a capable application worth keeping on your computer.

1 In the main window, go to the bottom left and set the sample rate to 44,100 kHz
2 Navigate to the menu bar and select Tracks – Add New – Mono Track

3 From the menu bar, go to Generate – Tone

   a Set Waveform to Sine
   b Set the Hertz to 100
   c Set the Amplitude is 0.8
   d Change the Duration to show Seconds
   e Set to one second
   f Click OK
   g Click the magnifying glass icon with a plus sign to zoom until you see the sine wave clearly
   h Press the space bar or the green play button to hear the sine wave

4 Go to Tracks – Add New – Mono Track

5 Go to Generate – Tone

   a Set Waveform to Square
   b Set the Hertz to 100
   c Set the Amplitude is 0.8
   d Change the Duration to show Seconds
   e Set to one second
   f Click OK
   g At the left of the square wave track, click the Solo button

      - This will mute the sine track so that you can hear the square wave on its own
      - Press the space bar or the green play button to hear the square wave
      - Press the Solo button again to unmute the sine track

6 Go to Tracks – Add New – Mono Track

7 Go to Generate – Tone

   a Set Waveform to Sawtooth
   b Set the Hertz to 100
   c Set the Amplitude is 0.8
   d Change the Duration to show Seconds
   e Set to one second
   f Click OK
   g At the left of the sawtooth wave track, click the Solo button

      - This will mute the sine and square tracks so that you can hear the square wave on its own
      - Press the space bar or the green play button to hear the square wave
      - Press the Solo button again to unmute the sine and square tracks

8 Go to Tracks – Add New – Mono Track

9 Go to Generate – Noise

a Set Waveform to Pink
b Set the Amplitude is 0.8
c Change the Duration to show Seconds
d Set to one second
e Click OK
f At the left of the noise track, click the Solo button
- This will mute the other tracks so that you can hear the noise on its own
- Press the space bar or the green play button to hear the noise
- Press the Solo button again to unmute the other tracks

10 Go to Tracks – Add New – Mono Track

11 Go to Generate – Noise

a Set Waveform to White
b Set the Amplitude is 0.8
c Change the Duration to show Seconds
d Set to one second
e Click OK
f At the left of the noise track, click the Solo button
- This will mute the other tracks so that you can hear the noise on its own
- Press the space bar or the green play button to hear the noise

12 Press the Solo button again to unmute the other tracks

13 If you want to save your project, go to File – Save Project – Save Project As

a Audacity will warn you that this will save the Project and not the audio file
b Press OK to proceed.
c Select the location where you want to save the Project
d Click Save

This concludes this activity. If you want, you can save the project or quit without saving.

Terms

Amplitude The volume or loudness of a sound waveform, measured in decibels.

Bandwidth A range of frequencies that a device can produce or capture.

Complex Waveform Any naturally occurring sound created by acoustic instruments, voices, and non-mechanical devices. Complex waveforms have random occurring harmonics.

Decibel The unit of measurement for the loudness of a waveform.

Envelope The change of the volume or loudness of a sound waveform over time.

Frequency The number of cycles per second that a sound waveform produces, measured in hertz.

Fundamental The initial or original tone of a sound, the loudest element in a sound.

Harmonics Frequencies occurring above the fundamental of a sound.

Hertz The unit of measurement for the frequency of a sound waveform. Abbreviated as Hz.

Longitudinal wave A waveform that moves forward and backward.

Phase The point of origin of the cycle of a waveform.

Pink Noise A waveform consisting of all frequencies playing at once but getting quieter at each octave.

Pitch The musical note that a sound produces.

Sawtooth Waveform A waveform consisting of a fundamental and odd and even harmonics.

Simple Waveform Any artificial sound created by a mechanical device. Simple waveforms have predictable and consistent harmonics.

Sine Waveform A waveform consisting of a fundamental and no harmonics.

Sound Pressure Level (SPL) The amount of acoustic or sound pressure in the air. An accurate measurement of sound with respect to hearing.

Square Waveform A waveform consisting of a fundamental and odd harmonics.

Transverse waveform A waveform that has an up-and-down motion.

Wavelength The physical length of a waveform if it is a straight line.

White Noise A waveform consisting of all frequencies playing at random volume levels all at once.

Chapter 3

Digital Audio

Digital Audio

Now that we understand sound waves and how they travel, we can discuss how sound is digitized and stored onto your computer. We must also look at how digital information is converted to acoustic energy that we hear through our speakers or headphones. Explaining digital audio can get very technical and mathematical. My goal is to cover the basic process of digital audio recording so that we understand how to work with it when creating our own music.

Before Digital

Before digital recording existed, the sound was recorded onto magnetic tape. This process became known as analog recording once digital recording emerged. We would use a microphone to capture sound. The microphone converted the motion of the air molecules into electrical energy, which ran down a cable to a mixing console. The mixing console allows an engineer to increase the volume of the sound and then send it down another cable to a magnetic tape recorder. The magnetic tape recorder took the electrical energy and converted it to magnetic energy. The magnetic energy was then stored linearly onto a magnetic tape.

Magnetic tape recording did have issues. Editing audio meant physically cutting the tape which could lead to errors. The tape would wear down a little each time it passed over the machine heads, so the quality of the audio degraded over time. The process of converting electrical energy into magnetic energy introduced noise in the audio often heard as hiss. However, magnetic tape recording was the sole method of recording audio from the 1950s until digital recording became an option in the early 1990s. Today, only a handful of studios still record to tape.

Digital Recording

Early digital recording still involved recording audio onto tape. Although the music was recorded onto tape, the data was stored as 1s and 0s, or digitally.

DOI: 10.4324/9781003345138-4

As a result, the tape was less prone to quality loss over time. The process of digital recording nearly eliminated hiss from the recordings. Hard disk recording emerged in the early 2000s, eliminating the need for digital tape decks. Hard disk recording stored digital audio onto a computer hard drive. You no longer had to fast forward or rewind the tape, you could specify exactly where you want to playback audio and the hard drive could jump to that location instantly. Hard disk recording meant you could easily edit audio and like a word processor, you could undo a mistake. Hard disk recording also brought digital recording to the home studio at an affordable price. A small investment allows you to record digitally anywhere you like and the ability to edit your recordings.

Digital Recording Process

Digital recording begins the same way as analog recording. We start with a sound wave, which is acoustic energy moving through the air. The microphone captures the acoustic energy. The sound wave causes the diaphragm inside the microphone to vibrate, which creates an electrical signal that matches the acoustic energy. The electrical signal travels along a cable and into your audio interface. The audio interface analyzes the audio and uses an analog-to-digital converter (ADC) to transform the electrical energy into digital data in the form of 1s and 0s. The digital data is sent to the computer where the digital audio workstation software application processes the audio and stores it onto your computer's hard drive.

When it is time to listen to the audio you recorded onto your computer, you press the play button on your software, and the computer sends the 1s and 0s back to the audio interface. The audio interface uses a digital-to-analog converter (DAC) to convert the data into electrical energy. The electrical energy travels from the audio interface to your speakers. The speakers transform the electrical energy into magnetic energy which moves the cone inside the speaker. The cone displaces air molecules which turns into sound. The entire process happens nearly instantaneously. If you take a moment to think about how quickly this all happens, you can appreciate the process.

Digitizing audio requires that the frequency of the sound and the amplitude is captured. In an analog system, the sound is converted to electric or magnetic energy. Both forms carry the frequency and amplitude in a single continuous stream. Digital systems capture the frequency of the sound as one variable and the amplitude as a separate one.

Sampling

The ADC must capture the pitch or frequency of the sound and its volume. Unlike magnetic tape recording, which captures audio as a continuous stream of information, digital recording requires capturing a snapshot of the frequency and loudness of the audio at regular intervals. The process of taking regular snapshots of audio is called sampling (Figure 3.1).

The issue with sampling is determining how many snapshots are needed to gather enough information to accurately represent the audio. A low-resolution photo of a flower will look blurry and pixelated. There is enough information for our eyes to interpret the image as a flower, but the image is not an accurate representation of the flower. A higher-resolution image shows more detail, and we can see the subtle changes of color in each petal. The same is true with audio. There might be enough data for us to recognize a sound as a flute, but if the sound is distorted or noisy, it is not an accurate representation of the instrument.

It is logical to assume that the more samples we take of an audio, the better the quality of the audio. With images, higher resolution images look better, but they also require more storage space on your computer. Images saved in a Tag Image File Format require more space but capture images in greater detail. These higher-resolution images take longer to load and require more processing power from your computer. An audio file made up of millions of samples may sound better but requires a large amount of disk space and processing power for the computer to accurately play back the audio.

Digital images use dots or pixels to capture color information. The resolution of a digital image is determined by how many dots are used per inch (DPI) when digitally capturing the image. An image captured at 300 DPI will look sharper and more vibrant than an image captured at 72 DPI. In audio, the resolution of digital audio is determined by how many samples per second are captured. This is called the sampling rate. The sampling rate is measured in frequency, or hertz (Hz). A sound with a sampling rate of 300 Hz means that 300 samples were captured in one second.

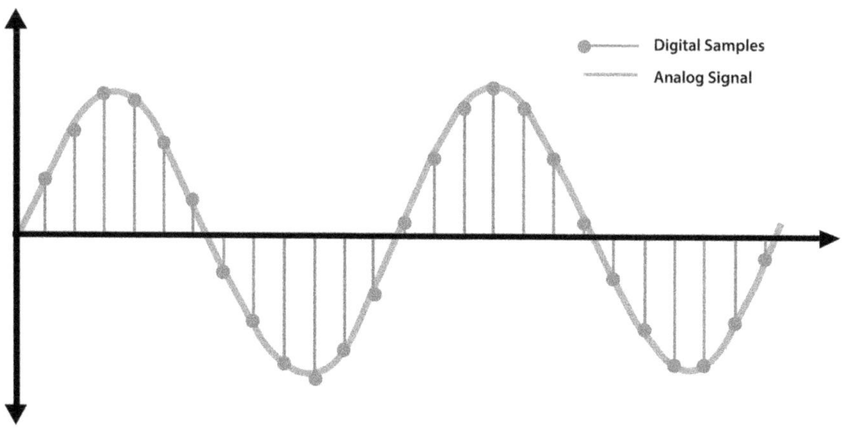

Figure 3.1 Digital sampling of analog waveform

We now understand that we take samples of audio every second to convert it to digital information. However, we need to determine the minimum sampling rate to accurately capture audio.

Nyquist Theorem

Harry Nyquist, a Swedish physicist and electronic engineer, calculated that to accurately capture a waveform we must take at least two snapshots per hertz of the highest frequency we wish to capture. If we have a waveform that is three hertz, then we need to take at least six snapshots. The reason is that we must capture the upper (positive) and lower (negative) portions of the waveform. If we take only three snapshots, then we are only capturing one portion of the waveform. Doubling the snapshots allows us to capture the positive and negative values of the waveform. You can see the results in the image given in Figure 3.2.

We know that the range of our hearing, or bandwidth is 20 Hz–20,000 Hz (20 kHz). If we want to sample all the frequencies within that bandwidth, then we need to take two snapshots of the highest frequency we wish to capture. To capture all the frequencies we can hear, we must capture all frequencies up to 20 kHz. According to Nyquist, we must take at least 40,000 samples per second to accurately capture that range. An ADC must use a sampling rate is 40 kHz, twice the desired frequency of 20 kHz, to accurately capture sound.

ADCs are simple devices that attempt to process all sound that they encounter. If the ADC can sample up to 20 kHz accurately, what happens if a sound at 21 kHz enters the system? The ADC will attempt to sample the

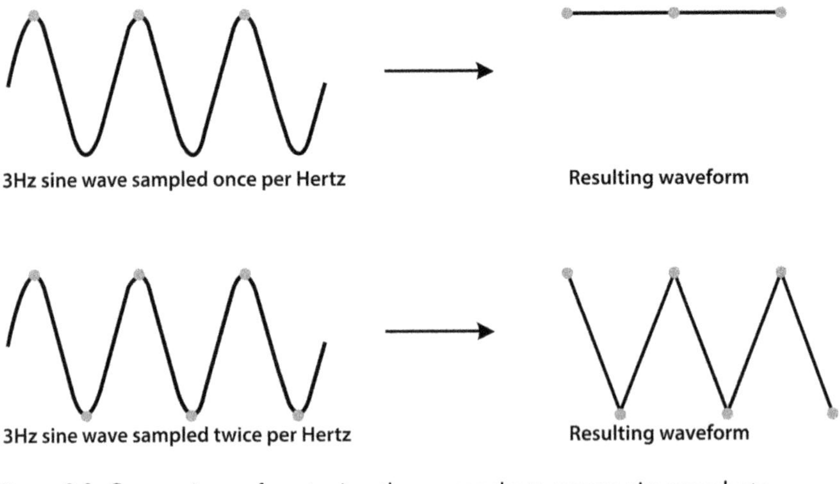

Figure 3.2 Comparison of capturing three snapshots versus six snapshots

frequency but will encounter a mathematical error when it discovers that it cannot generate enough samples to capture the frequency. This error will generate an audible noise or distortion in the audio stream called aliasing.

Aliasing

Aliasing occurs whenever an ADC cannot distinguish or differentiate one frequency from another. An ADC with a sampling rate of 40 kHz can only capture frequencies up to 20 kHz. When a 21 kHz sound enters the system, the ADC will assume it is a 20 kHz sound since that is the highest it can sample. The resulting sample is inaccurate and causes an error. To prevent such errors from occurring, all frequencies above 20 kHz must be removed before the sound enters the ADC.

The process of removing frequencies from a sound is called filtering. When we filter something, we remove the items we do not want, so we see only what we want. If you search for an audio interface on a website, you can filter your search to only show USB interfaces. The same process applies to sound; we want to filter the sound so only frequencies up to 20 kHz are processed. We need all the frequencies below 20 kHz to pass. The filter we use is called low-pass filter. A low-pass filter allows only frequencies below a specified frequency to pass. In this scenario, we would set the low-pass filter at 20 kHz.

Quantization

The sampling rate of an ADC captures the frequencies of audio. Now we must look at the process of capturing amplitude. Audio enters the ADC as electric energy. The voltage of the electric signal determines the amplitude of the audio. The greater the voltage, the louder the sound. We must also remember that the voltage of audio is continuous and constantly changes. The ADC must quantize the voltage of the audio on a scale every time a sample is taken. If we have a scale from one to ten, then we could say that this first sample has a value of two while the second has a value of four, and the third has a value of six. If the voltage falls on a line when it is measured, then we have an accurate measurement. Any amplitude that falls between two lines would need to be rounded to the closest line. Rounding an amplitude level up or down to the closes level creates a quantization error. This error results in noise or hiss in the recording.

To limit the number of quantization errors, we need to increase the number of levels we measure against. The measured level is stored as a binary number with a fixed number of digits. The number of digits determines the word length or bit depth of the audio. An 8-bit word has eight pairs of binary numbers. The number of levels is calculated using the formula two to the power of n, where n is the bit depth. In the case of our 8-bit word, two to the power of eight results in 256 levels. To put it another way, an 8-bit ADC can measure up to 256 levels of amplitude per sample.

Professional standards require a minimum bit depth of 16 (65,536 levels), although many prefer a bit depth of 24 (16,777,216 levels). The increased number of levels significantly decreases quantization errors, but also determines the dynamic range of the ADC. Dynamic range is the range in decibels (dB) between the lowest and highest amplitudes that a device or ear can capture or produce. Using 120 dB as the threshold of pain for our hearing, we can say that our dynamic range is 120 dB. When sampling audio, the higher the bit depth, the higher the dynamic range. Audio at 16 bits has a dynamic range of 96 dB while 24-bit audio has a dynamic range of 144 dB.

It is easy to assume that 24-bit audio would sound louder than 16-bit audio because the dynamic range is higher. With analog systems, like our ears or a speaker, dynamic range measures the decibel level at which the ear of device experiences pain or damage. A speaker with a dynamic range of 120 dB means that it can withstand sound up to 120 dB in amplitude before distorting or being damaged. With digital systems, a higher bit depth means more levels of measurement, which lowers the number of quantization errors. Quantization errors create noise or hiss in the audio, so fewer quantization errors mean less hiss. Lower noise or hiss increases dynamic range (Figure 3.3).

Noise Floor

Noise floor is any sound or noise that prevents you from hearing extremely quiet sounds. A busy intersection can produce 75 dB of noise. If someone is

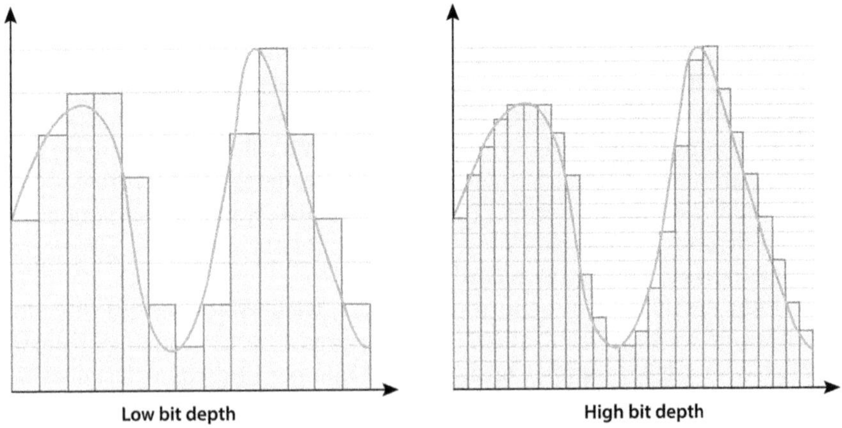

Figure 3.3 A sine wave captured at different quantization levels

speaking to you at 50 dB, you will not hear them. Subtracting the noise floor from your dynamic range yields your actual dynamic range. Standing at the busy intersection reduces our dynamic range to 45 dB (120 – 75). A speaker may have a dynamic range of 120 dB but if the electronics inside the speaker produce 20 dB of noise, then the actual dynamic range of that speaker is only 100 dB.

The magnetic tape recording process creates tape hiss. This is unavoidable. Professional multitrack recorders have a noise floor of 30 dB while a cassette deck has a noise floor of 55 dB. If the music on the recording is loud, then the noise is less noticeable because the sound of the music can mask the hiss. The busy intersection has a noise floor of 75 dB but if an ambulance passes by at 110 dB, you will hear more of the ambulance and less of the street noise. However, if the recording is a classical guitar playing quietly, the tape hiss is pronounced and distracting.

Even with quantization errors, a digital recording has significantly less noise than a magnetic tape recording. Reducing the noise floor increases the dynamic range. A 24-bit recording with a dynamic range of 144 dB has a lower noise floor than a 16-bit recording with a 96 dB dynamic range. The difference between the two seems large, but if you consider that most of us experience at least 40 dB of environmental noise around us, our effective dynamic range is only 80 dB. Either bit depth exceeds our own dynamic range so hearing the difference between the two is unlikely (Figure 3.4).

Dither

When working with digital audio, you will come across the term dither. Dither is simply noise added to a digital recording. Given our discussion of noise and the advantages of digital recording, why would we add noise to our signal? Regardless of the bit depth, there will be quantization noise in our

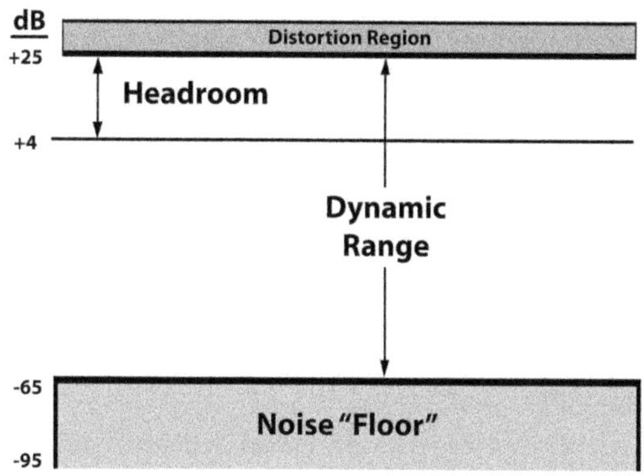

Figure 3.4 The effect of noise floor on dynamic range

recordings because not every amplitude level will fall exactly on a line. Dither is shaped noise that pushes the quantization noise to a higher or lower bandwidth, where it is masked by other sounds so that we do not hear the noise. The result is subtle and may not be perceptible by all listeners. Dither ensures that the noise is not in a noticeable bandwidth. Dither is not a requirement in digital audio; it is just a bit of assurance in preventing noise.

Digital Audio Playback

Playing back digital audio from your computer requires the audio interface to have a DAC. The DAC receives the digital stream and converts the data into electric energy that can be sent from the interface to your speakers. The DAC reconstructs the frequency of the waveform and assembles it sample by sample. The DAC then reads the quantization levels of each sample and recreates the voltage for the amplitude of the waveform. This entire process has quickly so that there is no noticeable delay between the time you press play and hear the sound.

The Audio CD

Digital recording first existed in the professional audio market. These early systems stored the data on digital videotape. It was not until the release of the digital audio compact disc (CD) in 1982 by Sony that digital audio reached the consumer. The CD stored audio digitally onto an optical disc. The disc could not be scratched or damaged as easily as a vinyl record. It also did not have the noise floor of a vinyl record or cassette tape. The CD was a joint development between Philips and Sony. The CD established the first digital audio standard, which is a sampling rate of 44,100 Hz and a bit depth of 16. Let us look at how Sony and Philips arrived at these numbers.

Sony set the sampling rate of the CD at 44.1 kHz out of practicality. In the 1980s, hard drives were expensive and small in capacity. They were not large enough to store the amount of data needed for digital audio. Sony was already using digital video recorders and decided to use that existing system to store digital audio. The video recorders that they used captured video at a rate of 30 frames per second. Each video frame required 490 lines of storage on the tape. Sony calculated that they could store three audio samples per line of data on the tape. This meant that they could store three times 490 (1,470) samples per frame. Since the tape stored 30 frames per second Sony was able to store 1,470 times 30, or 44,100 samples per second. Since Sony was already using this sampling rate for their digital videotape recordings, they kept it for the CD.

The bit depth of 16 was less of a practical decision and more of a personal one. Philips had developed 14-bit ADCs and DACs while the ones used by

Sony were 16-bit. Sony eventually won the debate and 16-bit became the standard. In case you were wondering, 14-bit audio has a dynamic range of 84 dB and 16-bit audio is 96 dB. This is not a significant difference.

The 4.7-inch diameter size of the CD is another standard that was due to personal preference. The original specification called for a four-inch diameter disc to hold approximately 60 minutes of music. The story tells us that the president of Sony at the time wanted the CD to store more music so that he could listen to the entire ninth symphony by Beethoven without interruption, which was 78 minutes. Vinyl records were limited to 22 minutes of music on each side, so listening to the symphony required the listener to flip the record once, then change the record, and then flip it again. Cassettes were limited to 40 minutes per side in length to ensure that the tape was thick enough to withstand multiple rewinds and fast forwards. Therefore, to allow the president of Sony to listen to Beethoven's ninth symphony uninterrupted, the CD size was increased to store 78 minutes of music.

Digital Audio Standards

Digital audio has different stands depending on the medium or project you are working on. You should be familiar with these standards in case you are asked to work on specific types of projects.

Consumer CD
44.1 kHz sampling rate
16-bit depth

Digital Video
48 kHz sampling rate
24-bit depth

Professional Audio
96 kHz sampling rate
24-bit depth

High Definition Video
192 kHz sampling rate
24-bit depth

The sample rate and bit depth that you work with in your home studio depend on your own preference. If you plan on using streaming services and digital downloads to distribute your music, then working at the consumer CD standard is acceptable. Your sample rate and bit depth settings also affect the amount of hard drive space you need to store the audio.

Digital Audio Formats

The digital audio process creates files that require large and fast storage systems as well as fast processors to process the data. The amount of storage needed for digital audio depends on the sampling rate, bit depth, and number of audio channels. Digital audio can be stored as an uncompressed or a compressed file.

Uncompressed File Formats

Uncompressed audio formats store audio without compromising or removing any data. All digital data is retained when the file is saved. Uncompressed audio can handle sample rates up to 384 kHz and bit depths up to 32 bits. Uncompressed audio can be saved as a .WAV or .AIFF file.

The .WAV file is the predominant file format for Windows computers. Standard .WAV files only store the audio data. A variation of this file type is the Broadcast Wave File, sometimes abbreviated as BWF. BWFs still use the .WAV file extension but allow metadata to be stored with the audio information. Metadata is additional information stored within the file that allows for indexing and tagging. Metadata includes information such as the composer, title, album, year, genre, and length of the file. Metadata also adds timecode information which is used to synchronize audio and video files in film production. The .AIFF file is the predominant file format for Mac computers. This format stores metadata and all audio data.

Sonically, there is no difference between the two file formats. Windows and Mac computers can read both file formats as well. This is helpful if you are working with someone on a Mac computer and you are on a Windows computer. Many audio applications will default to one format regardless of whether it is installed on a Windows or Mac computer. If you want to retain all the audio information you record, choose one of these two formats to save your files. You do need to consider file size when working with either format. The higher the sample rate or bit depth, the larger the resulting file. Table 3.1 shows the amount of storage you need for one minute of stereo audio depending on sample rate and bit depth in megabytes (MB).

Table 3.1 Storage requirements for different sample rates and bit depths

	44.1 kHz	*48 kHz*	*96 kHz*	*192 kHz*
16-bit audio	10 MB	11 MB	22 MB	44 MB
24-bit audio	15 MB	16 MB	32 MB	64 MB

Compressed File Formats

The primary goal of compressed audio formats is to reduce file size. Compressed file formats reduce the file size by removing or compressing information. These formats use algorithms that decide what audio information is important for you to hear. Information that is not important is removed from the audio. The more data that are removed, the smaller the file becomes. Compressed file formats also reduce the amount of data that is streamed through the ADCs and DACs. These variables can reduce an audio file so that it is 10% of its original size. Compressed file formats can be lossy or lossless.

Lossy Audio

Compressed audio file formats arose out of the need to stream audio over the internet at a constant rate without interruption. Compressed audio formats remove audio information to reduce the overall file size. Removing audio information must be done carefully. If too much information is removed, the listener will notice the lack of quality. Audio compression algorithms must be able to remove information without affecting the overall perception of the audio quality. If a listener does not perceive any difference in audio quality, then they will not be aware that the audio that they are listening to is compressed. The most popular compressed audio format is the MP3 format. The MP3 file format was developed in the early 1990s by the Fraunhofer Society in Germany.

Compressed audio operates on the principle of perceptual encoding. Perceptual encoding captures only the audio information that is critical to the listener. Perceptual encoding analyzes the sound and determines the important elements of the sound and then removes everything else. The decision on what to remove is based on the principle of masking. Masking is when one sound covers up your perception of another sound. If you are working in your room and you hear a conversation outside, you can increase the volume of your speakers so that you do not hear the conversation. In this scenario, you are using the music to mask the sound of the conversation. The perceptual encoding algorithm would assume that you would rather hear the music than the conversation and attempt to remove conversation data.

Our hearing is most sensitive between 2,000 and 5,000 Hz, where speech occurs. Perceptual encoding retains information within this frequency range so that we can hear the lyrics of a song clearly. We perceive lower frequencies and higher frequencies louder than we hear mid frequencies. Perceptual encoding can remove some of the low and high frequencies without affecting our perception of those frequencies. Frequencies toward the upper end of our hearing are also removed. Compressed audio formats generally contain very little information about 12 kHz. Compressed audio formats are for the general consumer. Audio engineers train their ears to distinguish frequencies precisely so they can make accurate recordings and mixes. An audio engineer is more likely to notice missing information in an audio file than a general consumer of music.

Lossless Audio

Lossless audio formats used enhanced compression schemes to reduce the file size without compromising audio quality. The file is lossless because there is no loss in audio quality. Rather than removing data, this format looks for similarities within the audio file and compresses that information. The audio is encoded with a set of instructions of how the file was compressed. When you play back the audio, those instructions help decode the audio and restore all the compressed data. The compression scheme can reduce the original file size by 50%–70% depending on the complexity of the music. Classical and film music featuring orchestras will reduce less than popular music. The most common lossless format is the Free Lossless Audio Codec, or FLAC. Apple has its own format called the Apple Lossless Audio Codec (ALAC) and Microsoft has a lossless format called Windows Media Audio (WMA).

Compressed File Sizes

The storage requirements for compressed audio formats such as MP3 vary depending on the format and the streaming or bit rate of the audio. Bit rate is different than bit depth. Bit depth describes the levels of quantization used to capture the amplitude of sound. Bit rate is the number of bits per second transmitted along a digital stream over the internet or on your computer. The streaming rate of compressed audio is measured in kilobits per second (kbps). A kilobit is 1,000 bits. An MP3 file with a bit rate of 128 kbps will stream faster and easier than one at 320 kbps.

Table 3.2 illustrates the difference in file size and bit rates for uncompressed and compressed audio formats. The table uses a one-minute stereo 16-bit WAV file at 44.1 kHz for comparison.

Table 3.2 Storage requirements and bit rates for compressed audio files

File Format	*Bitrate*	*File Size*
CD audio	1,411 kbps	10 MB
MP3	320 kbps	2.4 MB
MP3	256 kbps	2 MB
MP3	192 kbps	1.5 MB
MP3	160 kbps	1.2 MB
MP3	128 kbps	1 MB

Computer Performance

Digital audio requires a fast computer and fast storage to prevent errors and audio dropouts. Digital audio also requires storage space, especially if you are working on several projects with several tracks. Uncompressed audio formats require a processor fast enough to efficiently stream audio without interruption. The storage drives must be fast so that data can be pulled quickly for processing. The computer requires sufficient RAM to store audio data as it is temporarily moved from storage to memory. Compressed audio formats, whether lossless or lossy, are not as demanding as uncompressed formats. Therefore, it is possible to stream audio from a mobile phone as easily as from a computer.

Recording digital audio will always require more processing power than playing back the audio. Even though the ADC and the DAC reside on the audio interface, the computer must still manage the sampling and quantization process while writing information to storage. Playing audio back requires less processing power since data is being read and not written. The more tracks you record at once, the more demands are placed on the computer. If you plan on recording many tracks at once, do not compromise on the specifications of your computer.

Summary

In this chapter, we looked at the process of digitizing audio. We began with a short history of recording and the transition from magnetic tape to digital recording. We learned the process of sampling and quantization and what the requirements are for each so that audio is accurately captured. We discussed digital audio standards and how digital audio was delivered to the consumer through the compact disc. We explored uncompressed and compressed audio formats and their requirements and benefits. Finally, we considered the effect on computer performance when working with digital audio.

At some point, you will probably find yourself in a conversation comparing the quality of magnetic tape recording over digital recording, with the magnetic tape being the better of the two options. I have avoided this discussion because as someone who is starting out with audio, the advantages of digital audio outweigh analog recording. Recording in an analog format is still expensive and requires several components and physical space to execute. Digital recording can be completed on a laptop and audio interface at a fraction of the cost without any loss in audio quality. This fact alone makes digital recording appealing and practical.

Activity 3

Digital Audio Samples

In this activity, we will use Audacity to closely examine a waveform and see how digital audio represents audio. With Audacity, we can zoom in to a waveform far enough to see the sample points.

1. Launch Audacity
2. In the main window, go to the bottom left and set the sample rate to 44,100 kHz
3. Navigate to the menu bar and select Tracks – Add New – Mono Track
4. From the menu bar, go to Generate – Tone
    a. Set Waveform to Sine
    b. Set the Hertz to 440
    c. Amplitude is 0.8
    d. Change the Duration to show Seconds
    e. Set to five seconds
    f. Click OK
5. Click the magnifying glass icon with a plus sign to zoom in
    a. Keep clicking until you see the sample points
    b. You may want to drag the lower edge of the track to make the view taller
    c. You will see dots with lines extending above and below the center line
6. Notice the smooth shape of the sine waveform
    a. It is incorrect to believe that digital audio creates steps and that the steps mean that digital audio is inaccurate
    b. Each point allows for smooth lines to be created
7. Click the X at the top left of the audio track for the Sine wave to delete the track and waveform
8. We are going to look closely at a sawtooth waveform
9. Go to Tracks – Add New – Mono Track
10. Go to Generate – Tone
    a. Set Waveform to Sawtooth
    b. Set the Hertz to 440
    c. Amplitude is 0.8
    d. Change the Duration to show Seconds

   e Set to five seconds
   f Click OK

11 Click the magnifying glass icon with a plus sign to zoom in
   a Keep clicking until you see the sample points
   b You may want to drag the lower edge of the track to make the view taller
   c You will see dots with lines extending above and below the center line

12 Notice the sharp angles of the sawtooth waveform
   a Digital audio is able to capture smooth curves and sharp angles
   b Even at 44.1 kHz, audio is accurately represented

13 Let us look at a music file to see how it is represented digitally

14 Download the audio file titled Phenom from the companion website
   a Your computer will probably place it in your downloads folder

15 Click the X at the top left of the audio track for the Sine wave to delete the track and waveform

16 Go to File – Open
   a Navigate to the Phenom audio file
   b Select the file and press Open

17 Zoom in on the file until you can see the sample points
   a This is a stereo file so there are two lanes of audio to look at
   b Notice that the audio lanes are similar but not identical
   c Some of the waveforms change abruptly, but overall, the lines are smooth

18 Let us see what effect exporting the audio file has on the look of the waveform
   a In the menu go to File – Export – Export as MP3
   b Set the Bit Rate Mode to Constant
   c Set the Quality to 128 kbps
   d Set the Channel Mode to Stereo
   e Click Save
   f You can leave the metadata fields empty for now
   g Click OK

19 From the menu go to File – Open

   a Select the MP3 file
   b The file will open in a new window in Audacity

20 We know that the audio quality of an MP3 file is approximately one-tenth of a WAV file

   a However, zooming in to view the samples reveals no visual difference between the two files
   b At the sample level, an MP3 file looks identical to a WAV file
   c The difference between the two files can be heard, not seen

This concludes this activity. If you want, you can save the project or quit without saving.

Terms

Audio Interchange File Format (AIFF) Default uncompressed audio file format for MacOS computers. Can store metadata.

Apple Lossless Audio Codec (ALAC) Audio lossless audio compression codec developed by Apple for compressed audio.

Aliasing Distortion or artifacts created by digital audio systems when two frequencies cannot be distinguished from each other.

Analog to Digital Converter (ADC) An electronic device that converts audio in electrical energy format to digital audio through the process of sampling.

Bit Depth The resolution of number of volume levels captured during the sampling stage in digital audio conversion.

Bit Rate The number of bits per second transmitted by a digital system. Used to measure the compression rate of compressed audio formats.

Compact Digital Disc (CDD or CD) Optical storage format for digital audio developed by Philips and Sony.

Digital to Analog Converter (DAC) An electronic device that decodes digital audio information and converts it to electrical energy.

Dither Noise intentionally added to a digital audio stream to minimize quantization errors and control the noise floor of an audio signal.

Dots Per Inch (DPI) The measurement of digital images that determines how many dots of image information can lie within a square inch. The higher the DPI, the sharper the image appears.

Dynamic Range The range in decibels between the quietest and loudest sound that a device can produce or receive without distortion.

Free Lossless Audio Codec (FLAC) A free lossless audio compression codec for compressed audio.

Lossless Audio A compression format that reduces the size of an audio file with no loss in the audio quality.

Lossy Audio A compression format that reduces the file size of an audio file by removing unessential audio information.

Magnetic Tape Recorder An electronic audio recorder that converts electrical audio energy into magnetic energy and stores the information onto magnetic tape.

Metadata Additional data stored within an audio file that provides information about the audio file such as length, sample rate, bit depth, author, and song title.

MP3 A popular lossy audio compression codec for compressed audio. Can create audio files as small as one-tenth the original file size.

Noise Floor The level of natural or artificial noise in an acoustic space or electrical system.

Nyquist Theorem The theory that states to accurately capture a sound, the sampling rate must be a least two times the highest frequency captured.

Quantization The process of converting the volume of a sound in voltage to a binary number for digitization of the sound.

Sampling Rate The rate or frequency samples captured in a digital audio sampling system. Measured in hertz.

Waveform Audio File Format (WAV) Default uncompressed audio file format for Windows computers. The broadcast variant can store metadata.

Windows Media Audio (WMA) A lossy audio compression codec developed by Microsoft.

Chapter 4

Computers

Computers

The computer is the cornerstone of your home studio. Along with managing tasks such as operating system, web browsers, email applications, and productivity applications, the computer must now handle all audio operations within the digital audio workstation (DAW) application. The DAW application requires the computer to manage sound coming in and out of the audio interface, display the waveforms on the screen, perform editing tasks, and enable virtual audio effects and instruments. If you work on a project with video, the computer must keep the video synchronized with the DAW while performing all the other tasks. How a computer is configured determines how effectively and efficiently it will handle DAW applications.

The computer configuration you choose depends on the tasks you need the computer to complete. You do not necessarily need the most powerful and up-to-date system to complete your work. I know several composers and producers who run ten-year-old computers with older DAW applications who create and produce music without issue. They work within the limitations of their computers to complete their tasks. However, I should point out that when these individuals purchased their computers, they configured powerful systems to meet the demands at the time. They also continue to run older DAW applications because they are fully aware that the latest versions of the applications will probably not run on their computers.

Choosing and configuring a computer is easier now than it was ten years ago, but you still need to understand the function of the components and how they affect computer performance. Depending on your budget, your first DAW computer will be a compromise between power, function, and price. The goal of this chapter is to help you make informed decisions about the components you choose so that you can have a computer that performs the tasks required of the DAW application. Computer technology evolves quickly, and changes are unavoidable. Therefore, I will refrain from recommending specific brands or models, but rather focus on specifications and features.

DOI: 10.4324/9781003345138-5

Platform

The debate between Apple Macintosh (Mac) and Microsoft Windows (PC) has existed since DAW applications were developed. In the 1980s and 1990s, most DAW applications were written specifically for Mac computers because the Mac operating system allowed software developers direct access to specific hardware components. Windows PCs have since allowed software developers the same access and now nearly all DAW applications run on both platforms with the same feature sets and performance. The choice of what platform to use depends on you and your personal requirements. Let us look at the differences and similarities between the two platforms.

Apple Macintosh

Apple controls the hardware and software their computers use. Apple does this to ensure that there are no compatibility issues between the hardware and software. The Apple operating system, MacOS, is designed to closely integrate the hardware and software on the Macintosh (Mac) computer. As a result, Macs are stable and rarely show issues between the hardware and software. If the hardware you attach is compatible with the operating system, it will operate as expected. Apple maintains a high-quality standard for their computers; the build quality is excellent.

Macs, whether laptop or desktop, are sealed systems; you cannot open the hardware and change or replace components. The configuration you purchase for a Mac is fixed and cannot be upgraded later. You cannot install a larger storage drive or increase the amount of memory on the computer. You need to configure a Mac to have sufficient storage and memory to support your work for many years. Adding additional storage and memory will increase the cost of the computer, so you need to budget carefully.

Many users find the Mac operating system intuitive and unobtrusive. The MacOS operating system remains in the background allowing you to work without operation. Adding external hardware such as an audio interface or MIDI controller is simply a matter of plugging the device into an available USB port. In most cases, the drivers will install automatically requiring no intervention from the user. Apple computers are popular with those who work in all types of media, such as graphic design, film, and music.

Windows PCs

A PC (personal computer) is any computer running the Windows operating system. IBM used the term personal computer to distinguish the home computer they designed from a mainframe computer. Desktop and laptop PCs are available from several manufacturers. Desktops can be purchased as a complete system with all components installed and configured, or you can build your own custom PC and select all the components yourself. In either case, you have the option to add additional memory and storage after your purchase. Some models even allow you to upgrade the central processing unit (CPU) and graphics card. Laptops are configured by the manufacturer. Many PC laptops are sealed systems, but some manufacturers allow you to upgrade memory and storage as you can on a desktop.

The Windows operating system is stable and user-friendly. Windows grants users more access to system functions, which can lead to problems if one is not careful. A challenge for the Windows operating system is ensuring compatibility across different hardware manufacturers and combinations of hardware. Sometimes incompatible hardware combinations can lead to problems with the Windows operating system. Windows does require more user interaction when installing hardware, so you need to be comfortable with doing this work on your own.

There is no right or wrong decision when choosing between a Mac and a PC. Most of you already have a preference and will follow that preference. Both platforms are stable and support all the major DAW applications. Aside from some differences in keyboard shortcuts, once you are in the DAW application, all the functions and features of that DAW are identical, regardless of the platform.

Operating System

We know that Macs and PCs run MacOS and Windows operating systems. Most of us are familiar with operating systems on a basic level. We know that the operating system allows us to access the computer and provides us with an interface to interact with our applications. However, having a deeper understanding of an operating system can help us in the future should we run into problems with our computers. Without an operating system, your computer is simply a static collection of parts.

An operating system is a collection of programs that control the resources of a computer system. It is written in a low-level language that speaks directly to the CPU, memory, storage devices, graphics, and other components. The operating system is an interface between the users and the hardware. When a computer is powered on, the operating system accesses code on the hard drive and begins to check the hardware components. After the hardware is checked, the operating system loads code into the main computer memory and soon you see the login screen. The operating system coordinates computer resources so that high-priority functions have immediate access to resources while low-priority functions run in the background. Processing functions are allocated based on need.

Computer functions are organized in layers. The hardware component is the lowest layer. The operating system sits above the hardware layer and provides instructions based on what the application requires. The application sits above the operating system and relies on it to communicate with the hardware. When the DAW application needs access to the audio interface, the application asks the operating system to pass the information to the hardware. The application receives instructions from the user. The operating system manages all requests and distributes resources to fulfill those requests.

An operating system requires updates to maintain stability and security. Regardless of the platform, always keep your operating system up to date. Keep in mind that an update is different than an upgrade. An update retains the version number of the operating system and adds new features and fixes. On a Mac, the operating system update could appear as going from version 15.2 to 15.3. On Windows, the build number will change. An upgrade introduces a new version number, going from 15.3 to 16 on a Mac or Windows 11 to 12 on a PC.

Upgrades should always be handled carefully. Upgrades are major changes to an operating system that can affect your applications and hardware. Always check that your hardware and applications are compatible with the new operating system version. You might discover that your current hardware is not compatible yet or that it is no longer supported by the new operating system. If your computer is running with no issues, then it is good practice to wait a few months before upgrading the operating system. This is critical if you are in the middle of a project with a deadline.

Computer Format

The decision on whether you should purchase a desktop or laptop computer depends on your budget and your workflow. If you want the flexibility to work in multiple locations, then a laptop is ideal. If you primarily work in one place and require multiple screens or additional external hardware, then a desktop is a better option. Desktop computers can accommodate multiple storage devices, graphics cards, and additional components. Expandability options vary between manufacturers, however. Laptop users can purchase a docking station as well, allowing them to add screens and external hardware when working from home.

Laptops tend to be less powerful than desktop computers. The main reason for this is space, power, and cooling. The faster and more powerful the computer processor, the more power and cooling it requires, which can add weight to a laptop. The keep laptops portable, manufacturers will choose less powerful processors and strive to find a balance between speed and portability. Battery life is also affected by the processor speed.

Certain manufacturers offer workstation laptops, which provide more power and features. Workstation laptops often have larger screens, up to 17 inches, the capacity for a powerful graphics card, and multiple storage options. These laptops can be as powerful as a desktop and still offer portability. However, workstation laptops cost more than traditional laptops and are significantly larger and heavier. If you intend to carry your laptop to many locations, this might not be a practical option.

For many, the deciding variable between a laptop and a desktop will be the cost. Laptop computers generally cost more, sometimes up to twice the cost of desktops. Laptops are more prone to damage if they travel to multiple locations. While most desktops can be upgraded after purchase, not all laptop computers can be upgraded. Always check the specifications before purchasing to determine your options. Let us take a closer look at computer components and then discuss the minimum requirements to run DAW applications to create music.

Central Processing Unit

The CPU is the computer's brain. The CPU processes all instructions received from the operating system and applications. With help from the operating system, the CPU communicates will all hardware components and send specific commands to each component. The CPU will run different processes at different speeds or priorities, based on the instructions from the operating system. If multiple applications are running at once, the CPU will attempt to run all of them at once. If resources begin to run low, the CPU will slow down to handle all the requests. A DAW application will work best when it is the only running application on the computer.

Windows Processors

Windows PCs have two processors to choose from, Intel and AMD. Both companies offer processors ranging from budget-oriented to performance-oriented, depending on your needs and budget. Both companies offer dedicated processors and processors with embedded graphics processors. Processors with embedded graphics processors handle graphic requirements without needing to purchase a graphics card. This feature is helpful on a laptop since a dedicated graphics processor adds to the cost and power requirements. With a desktop, you have the option of adding a dedicated graphics processor later. The physical design of the processor differs between both companies, however, both brands of processors can run Windows and are supported by DAW applications.

When comparing processors for Windows PCs, we generally look at the processor speed, which is measured in gigahertz (GHz). We can also compare the number of cores in a processor. Both Intel and AMD use individual cores on a processor to allow for a greater number of simultaneous processes. AMD processors have physical cores while Intel processors use a combination of physical and virtual cores. These slight differences make it difficult to compare an eight-core 4.0 GHz processor from both companies on published numbers alone. Instead, we must rely on benchmark tests, which measure the processor speed across several tasks. You can find benchmark results on the internet to compare processor performance.

Choosing between an Intel and AMD processor depends on personal preference and other influences. AMD processors are extremely popular with gamers while Intel is favored by those working in media. AMD processors tend to offer more value for money, and many manufacturers now offer models using both brands. Read reviews and talk to other users. Both brands are viable and reliable options for Windows computers. Intel holds a greater market share, but AMD has gained a lot of popularity in recent years.

Apple Processors

Apple manufactures its own CPUs for all its computers. All Apple processors integrate a graphics processor. The physical design of the processor differs dramatically from Intel and AMD. The Apple processor is designed exclusively for the MacOS operating system and is only available on Apple computers. The processor is tightly knit with the motherboard and memory, so upgrades are not possible for any of the components.

The architecture of Apple's processors makes it difficult to compare them to Intel and AMD. Apple's design allows the processor to be more efficient, and thus can run at lower speeds, which reduces heat and increases battery life. However, if you are planning on purchasing an Apple computer, then you are probably not concerned about how the processor compares to one in a Windows PC. For most users, the decision between a Mac and PC is not based on performance, but preference over the operating system environment.

Apple has several versions of its processor, and the choices depend on whether you are choosing a laptop or a desktop. The latest versions of the MacOS operating system are designed specifically for their processor, so the overall performance of the computer is responsive and quick.

Motherboards

Those of you planning on building your own Windows desktop PC need to consider the motherboard. The motherboard houses all the computer composes and provides connections to your storage devices. Your motherboard choice depends on the processor you wish to run. You can purchase a small motherboard that allows you to connect to a single graphics card and only offers two memory slots. Or you can choose a larger model that can accommodate multiple graphics cards and has four memory slots. Choosing a motherboard for your computer build requires research and additional knowledge that is beyond the scope of this book. The companion website lists resources you can consult for further reading.

Random Access Memory

Random access memory (RAM) is physical memory used to temporarily store information. RAM is volatile and requires power to operate. Once RAM loses power, the information is lost. RAM size is measured in gigabytes (GB). RAM is designed for temporary and quick access to information, and a portion of it will always be used by the operating system. When RAM resources run low, the operating system will use the main storage device as temporary RAM. All operating systems will allocate a portion of the main storage device as temporary RAM, called virtual memory as soon as the operating system loads. Virtual memory is not as fast or efficient as physical memory. DAW applications rely on physical memory for recording and playback. Therefore, it is important that your computer has sufficient RAM to run DAW applications.

Graphics Processing Unit

The graphics processing unit (GPU) processes video information and draws images on your screen. Graphic designers and video editors rely on powerful graphics cards to render images and videos quickly. DAW applications may look graphic intensive, but the meters, dials, waveforms, and other elements do not require a significant amount of graphics processing. This means that processors with embedded graphics processing can easily handle the graphics demands of a DAW application.

If you are working on a Windows PC, and you plan on working with video, either as a composer or sound designer, then your graphics processing needs could increase. Digital video is often compressed to save space. Playing compressed video on your computers involves decompressing the video information and reassembling the picture. This task is handled by the graphics processor, and if the graphics processor does not have enough power and resources, the video playback will not be smooth or sharp. A short video will play back correctly, but a longer video may not. In this scenario, a dedicated graphics card is required. Fortunately, the gaming industry has prompted AMD and Nvidia to develop a range of graphics processors to fit every need and budget.

The GPUs from AMD and Nvidia come in a variety of specifications. Consult the manufacturer's websites for specifications and designs when choosing one. I will close by adding that the graphics processing on the Apple CPU is managed differently resulting in excellent video processing.

Storage Devices

Storage drives range in size, but most drives are at least 500 gigabytes (GB) in size and can be as large as four terabytes (TB), which is 4,000 GB.

There are two types of storage devices to choose from. The first type, and the oldest, is called a hard disk drive (HDD) and uses magnetic storage on rotating platters. The platters are coated with a magnetic material that stores data in blocks. The platters spin as quickly as 7,200 revolutions per minute (rpm) and use an arm to access different sections of the platter. The entire mechanism resembles and turntable. HDDs are available in large sizes at low prices. The moving components in an HDD mean they can be easily damaged by sudden shocks, such as a fall. These drives connect to the motherboard using a data cable and require power. HDDs are non-volatile; data is retained even after power is removed. HDD drives are still viable storage options because of their lower cost per GB compared to flash storage drives.

Flash storage is the second type of non-volatile storage device. Flash drives differ from HDDs in that data is stored on memory cells instead of spinning magnetic platters. There are no moving parts in flash storage drives making them more resilient to shocks. The drives are silent and cooler. Read and write times for data are significantly faster on flash storage. Flash drives come in two different formats.

The first type of flash storage is the solid state drive (SSD). These drives share the same dimensions and connections as HDDs ensuring compatibility with all computers. Flash SSDs are the second type of flash storage. Flash SSDs resemble memory chips and plug directly into the computer motherboard. The direct connection to the motherboard allows for even faster access to data. Newer desktop and laptop computers ship with flash SSDs.

Storage Capacity

We should take a moment to discuss how storage capacity is measured on computers. Given that our entire musical existence relies on computers, storage capacity is something we must understand and be comfortable with.

A bit is the smallest unit of measurement in a computer system. A bit is a binary digit and can either be a 1 or a 0. A binary system groups digits in twos, whereas our decimal system groups everything in tens. This small distinction is important as we increase the size of our capacity.

A byte is the next unit, made up of eight bits. A byte is any combination of eight 1s and 0s. A single character of text on your computer, like the letter A, requires a single byte of computer memory. The word "cat," therefore, requires three bytes of storage.

The kilobyte (kb) is the next unit. In a decimal system, kilo means 1,000, but in a binary computer system, kilo means 1,024. Therefore, a kilobyte (kb) is 1,024 bytes. From this point on, each larger unit of measurement is 1,024 times the size of the previous measurement. The abbreviations after kilobyte are also capitalized.

1 byte = 8 bits
1 kilobyte (kb) = 1,024 bytes

1 megabyte (MB) = 1,024 kb
1 gigabyte (GB) = 1,024 MB
1 terabyte (TB) = 1,024 GB

Drive Formatting

Storage drives must be formatted before an operating system can read and write data to the drive. The primary drive, also called the system drive, uses the default format for the operating system. Other drives connected internally to the motherboard can also use the operating system's default format. The same is true for external drives. However, if you intend to use an external drive to share data between Macs and PCs, then the drive format must read and write access to both operating systems

The default format for Windows computers is called New Technology File System (NTFS). The default format for MacOS computers is called Apple File System (APFS). Both formats support all drive types and sizes. Windows computers cannot access APFS drives. MacOS computers can read from NTFS formatted drives but cannot write to them. Neither drive format is suitable for an external drive that must share data between both operating systems.

External drives can use the Extensible File Allocation Table (exFAT) format. Both Windows and MacOS operating systems can read and write exFAT formatted drives. This drive format supports large drive sizes. Both operating systems have storage drive formatting utilities that will allow you to format an external drive as exFAT.

Computer Specifications

Now that we understand computers and their components, we need to talk about the needed specifications for a DAW computer. These specifications apply to both PCs and Macs.

Processor

The minimum processor requirement is a multi-core model with at least four physical cores. The processor can have additional physical or virtual cores. Intel and AMD processor speeds should be at least three gigahertz. Apple does not publish their processor speeds and all their processors have at least eight cores. The processors included with base model Apple computers have sufficient processing power.

Memory

RAM is critical to the performance of a DAW. Most DAW applications state that eight gigabytes of RAM is the minimum requirement. Remember that when your computer runs low on RAM, it must use virtual memory on the storage drive, which is must slower than RAM. For these reasons, you should choose a computer with at least 16 GB of RAM. This will increase the cost of the computer, but the extra memory will improve the performance of your DAW.

Graphics

The graphics processor on Apple computers depends on the selected processor. PCs with embedded graphics processors on the CPU can handle DAW graphics efficiently. If you plan on working with video, then a dedicated graphics card is recommended. PC laptops may offer you an option for a dedicated graphics card, whereas desktops always have the option for a dedicated graphics card. Choose a graphics card with at least four gigabytes of video RAM (VRAM) for smoother video playback. Both AMD and Nvidia offer excellent solutions.

Storage

A DAW computer should have at least two hard drives; one for the operating system, the other for audio data. The first drive stores the operating system and all your applications. The second drive stores audio files. A dedicated audio drive improves the performance of your DAW because the data can stream from the drive without being interrupted by operating system tasks. Desktop computers can have two internal drives. Ideally, both drives should be flash storage drives. Some laptops allow for two drives, but most do not. In this case, an external drive is the best solution.

The system drive should be at least one terabyte (TB) to accommodate the operating system, application, and sound libraries. The audio drive should be at least one terabyte in size as well. Larger drives will ensure that you do not run out of space when working on several projects.

Compromises

Your budget may require you to compromise on your selections. Here are some suggestions to help you decide.

Memory over processor speed. Memory is critical to DAW performance since all data passes through RAM. If you must choose between a fast processor and more RAM, choose more RAM.

Processor over storage. If adding a second drive exceeds your budget, then remove it from your specifications. You can add an external drive later.

Everything over graphics processor. The graphics processor should be the last item on the list. It is nice to have a dedicated processor, especially when working with video, but you can always reduce the quality of the video if it starts to affect the computer's performance. The image may not be as sharp or clear, but sufficient to complete the project.

Computer Connections

Desktop and laptop computers have several ports that allow you to attach additional devices. Some of these ports can accept a variety of devices while others are specific to devices.

Universal Serial Bus Ports

The Universal Serial Bus (USB) protocol is ubiquitous, and every computer has at least a couple of USB ports. USB ports can accept any number of devices such as printers, keyboards, mice, scanners, audio interfaces, MIDI controllers, and even microphones. If your computer does not have enough ports, you can purchase a USB hub to increase the number of ports available on your computer.

The USB-C connector is more common now on newer computers. USB-C connections are the same for the host computer and the device. Prior to USB-C, the computer had one type of USB connector while the device might have had a different one. The connections became even more complicated when micro and mini-USB connectors were added to smaller devices. This meant that you needed to have a collection of different USB cables to accommodate different types of ports. The simple design of the USB-C connector and cable means that a single cable can be used to connect any USB-C device to a computer. USB-C ports are backward compatible allowing you to connect a device with an older connection provided you can the correct cable.

Thunderbolt Ports

Thunderbolt is a protocol that provides devices with direct connections to your motherboard for extremely fast transfer rates. A Thunderbolt device operates at the same speed as if it were connected to an expansion slot in your computer. Thunderbolt supports display connections to your computer monitor, external hard drives, external graphics card, and audio interfaces. Thunderbolt devices can be daisy-chained without compromising transfer speeds or performance. A single Thunderbolt connection can support up to six devices. This means you could connect a computer monitor, audio interface, and hard drive to one port by daisy-chaining the devices.

Thunderbolt connections are standard on Mac computers and only found on a limited number of PCs. Thunderbolt is currently in its fourth generation and shares the same size connector as USB-C. To connect a Thunderbolt device to this port you must use a Thunderbolt cable. Thunderbolt connections are backward compatible with USB-C. You can connect a USB-C device to a Thunderbolt connection, and it will operate as a USB-C device. Since both protocols use the same port, the only way to distinguish them is by the lightning bolt symbol next to the port on the computer and the same symbol on the cable.

Display Monitor Ports

To connect a display monitor to your desktop or laptop you need one of three connections. The High-Definition Multimedia Interface (HDMI) is common on consumer devices such as televisions. HDMI supports high resolutions and can carry audio signals as well.

DisplayPort (DP) is a specialized port that also supports high resolutions and audio but at much higher rates, meaning that images refresh faster on your screen. DP is popular with gamers because the chances of video lag are less when using an appropriate graphics card.

I mentioned earlier that you can connect a computer display to your Thunderbolt port. Thunderbolt uses the DisplayPort protocol for video signals.

All three protocols are compatible with each other. You can connect an HDMI monitor to a DP or Thunderbolt connection using an adapter. Older monitors will only have HDMI connections whereas newer models have HDMI and DP connections. The type of cable you need depends on the port on your desktop or laptop.

Device Drivers

When we connect a hardware device such as an audio interface to a computer, the operating system needs information to communicate with the device. A device driver provides the operating system with the code and files needed to communicate with the hardware device. Without the device driver, the operating system cannot communicate with the device, rendering it unusable. The device driver tells the operating system what the hardware is, what it does, and how to use it. The device driver is provided by the manufacturer, although in many cases, it is already included with the operating system.

If the device driver is included with the operating system, then the device will be automatically recognized and configured upon connection. Devices that install automatically are called class compliant. Class-compliant hardware does not require manual driver installation because the operating system already supports the device. Devices such as keyboards, mice, printers, external drives, and webcams are often class compliant.

Manually installing a device driver requires you to run an application that will guide you through the installation. This application will need to be downloaded directly from the manufacturer's website. Doing so ensures that you always have the latest driver for your device. Make sure the driver you download is the correct one for your operating system and version. Some devices will require you to reboot the computer while others will not. On rare occasions, you might need to install the driver before connecting the device. Always read the included documentation for the device before connecting it to your computer.

Audio Device Drivers

Audio device drivers require special attention because they differ between MacOS and Windows operating systems. The MacOS operating system uses a single audio driver for all audio applications. This driver is called Core Audio and is included with the operating system. The Core Audio driver manages the system sounds of the computer such as warning beeps, audio and video playback, web browsers, and other system-level sounds. The Core Audio driver is a professional-level driver, suitable for the demands of a DAW application. The driver is multi-client, meaning you can watch a video on the internet while playing audio from your DAW simultaneously without compromise.

All Mac-compatible audio interfaces support Core Audio. When you connect an audio interface, the MacOS operating system automatically installs the driver. The Core Audio driver provides basic functionality for the audio interface. Some manufacturers include an enhanced driver and utility that allows you access advanced features within the audio interface. The audio interface will function fine without the enhanced driver, but you may be missing out on additional features and options.

Audio drivers on the Windows operating system, on the other hand, are a bit more complicated. The Windows Driver Model (WDM) handles the basic needs of the operating system. Windows uses the WDM audio driver to manage audio for system sound effects, audio and video playback, web browsing, and other system-level sounds. The WDM audio driver is designed to work with the built-in sound card included on Windows desktops and laptops. The WDM driver is flexible, but it is not designed for the real-time demands of a DAW.

To overcome the inefficiency of the WDM audio driver, Steinberg, the manufacturer of the DAW application Cubase, developed a driver specifically for professional audio called Audio Stream Input/Output (ASIO). The ASIO driver supports real-time audio processing and professional audio fidelity. ASIO only supports professional applications, meaning that Windows cannot access it for system-level sounds. This is not an issue since the built-in sound card already handles those needs with the WDM driver. When you connect an audio interface to a Windows computer, you must install the ASIO driver before the DAW will recognize your interface.

Applications

An application is a set of software codes designed to perform a specific task. Applications rely on the operating system to provide the framework it needs to function. The application needs the operating system to provide access to the CPU, graphics processor, system memory, keyboard, mouse, and other devices. The application may also ask the operation to allocate specific resources. Without the operating system, the application cannot function.

Installing Applications

Both operating systems require you to install applications to the system drive. The system drive is where your operating system resides. Installing applications can differ between MacOS and Windows. The MacOS operating system allows applications to be installed in one of two ways. The first way is in the form of a package. A package is a single file that you drag in the Applications folder on the system drive. The package automatically extracts itself and writes all the required files to the Applications folder. Other MacOS applications will use an installer that will guide you through the installation with a series of prompts.

Applications written for the Windows operating systems use an installer with prompts to install the required files. The design of the Windows operating system often requires that the installer place files in the Program Files folder and several other locations.

In many cases, the application installer will require you to reboot your computer before you are able to use the application.

Removing Applications

Applications on both operating systems must be removed or uninstalled so that all application files are removed from the operating system, regardless of where they are located. Windows applications must be removed using the uninstaller included with the application. This is often done through the Control Panel or Settings in Windows under Programs and Features. Many MacOS applications can be removed simply by dragging the application out of the Applications folder and into the Trash. Some applications will require you to use the uninstaller application found inside the specific application folder.

Plug-Ins

A plug-in is a software application that runs on top of an application. A plug-in adds features or functions to an application. A plug-in requires a host application to function; it cannot function on its own. The plug-in code must be compatible with the application for it to function. If an application does not support the plug-in code, the application will not load the plug-in. If you have multiple applications that support the plug-in, those applications can load the plug-in. Plug-ins have their own settings, separate from the application. Some applications have their own library of settings so that you can recall the settings on multiple host applications. Others will save the settings with the host application. Plug-ins are generally small and do not require much storage space.

There are three plug-in formats for audio applications. There is no difference in how the plug-in operates in any of the three formats. The formats depend on the operating system and host application. Not all plug-in formats work with all operating systems and host applications. Most plug-in manufacturers develop their plug-ins to work in all three formats. Always check the plug-in specifications to ensure that the product will work with your audio applications.

The Audio Unit (AU) plug-in format, developed by Apple, is supported only on Mac computers. Most DAW applications that run on MacOS support AU plug-ins. The Virtual Studio Technology (VST) plug-in format, developed by Steinberg, is supported by Mac and Windows computers. All cross-platform DAW applications support VST plug-ins. The VST standard is currently on version three, which is named VST3. The Pro Tools DAW application, developed by Avid, supports its own plug-in format called Avid Audio eXtension (AAX). The AAX plug-in format is only available within the Pro Tools application. This plug-in format is supported on both Windows and Mac OS computers.

Every DAW application ships with a selection of plug-ins. The selection consists of audio effects and virtual instrument plug-ins. Audio effect plug-ins include equalization, compression, chorus, delay, reverb, and other effects to change the way audio sounds. Virtual instruments are software representations of synthesizers and other instruments you can use instead of recording acoustic instruments. The selection varies among manufacturers, but it is worth taking the time to explore the choices included with the DAW applications. If the selection in your DAW does not meet your needs, there are several companies that focus exclusively on designing audio effects and virtual instrument plug-ins.

Summary

Understanding a computer and its components is essential when using DAW applications. Our workflow depends solely on the performance and reliability of our computers. If we can learn to solve our own computer problems, we can save time and money. In this chapter, we covered the basic components of a computer system and how they interact with each other. We discussed the differences between the two main operating systems available to you and how each operating system handles plug-in formats. We learned the minimum computer specifications needed to run DAW applications. If you wish to increase your knowledge of computers, there are several resources on the internet where you can learn more about processors, motherboards, and graphics cards.

Activity 4

Computer Recommendations

I have created this computer recommendation list to help you determine your computer needs. Although the purpose of the recommendation is to help you narrow down your choices, please keep in mind that you still have the freedom to choose the system that is best suited for your workflow, regardless of the recommendations.

I also want to add that these recommendations assume you are purchasing a fully configured computer. Building your own computer is beyond the scope of this book. For many of you, purchasing a configured computer will be much easier, especially if you are just starting out with music production. The recommendations provide you with minimum and preferred requirements as well as additional suggestions depending on your workflow.

Recommendations

Operating System
Choose the operating system you are most comfortable using
MacOS
Windows

Platform
Laptops offer portability while desktops are easier to expand
Laptop
Desktop

Processor Windows
Windows – Intel or AMD
4-core minimum
8-core or higher recommended

Processor MacOS
MacOS – Apple M Series
8-core minimum
10-core or higher recommended

Memory (RAM)
8 GB minimum
16 GB preferred for more workloads
32 GB if working with more than 16 tracks of virtual instruments

Storage – Single drive systems
1 TB SSD minimum
2 TB SSD if working with virtual instruments and sample libraries

Storage – Multiple drive systems
1 TB SSD for first drive
1 TB SSD for second drive
2 TB SSD for second drive if working with virtual instruments and sample libraries
Add a third drive if composing orchestral scores for additional sample libraries

Graphics Card – MacOS
Desktops and laptops
Included in the processor, no options

Graphics Card – Windows
Desktops and laptops
Intel or AMD integrated graphics processor minimum
Nvidia or AMD GPU preferred if working with video
4 GB VRAM minimum

Ports
Multiple USB 3.0 or higher ports preferred
Laptop users should consider a powered USB hub if connecting multiple hardware
Thunderbolt 3 or higher optional for future expansion

Laptop Screen
Laptop screen sizes vary between 14 and 16 inches
Larger screens add weight
Minimum 1080p resolution

Desktop Screen
21-inch screen minimum at 1080p
24 inches or larger if using higher resolutions
Both laptops and desktops can use two screens if needed

Keyboard and Mouse
A keyboard with a numeric keypad is recommended
Many DAW applications use the numeric keypad for additional shortcuts
Laptop users can add a USB numeric keypad if needed
Any mouse that you are comfortable with will work

External Storage
An external storage drive is strongly recommended for backups

Conclusion

This list is meant as a starting point for choosing your computer specifications. Many of you may already have computers that meet the recommendations already, in which case, you are ready to start working. Some desktop users may be able to upgrade components to improve their specifications. Laptop users should try to purchase more than they initially need since most models cannot be upgraded after purchase. As always, work within your needs and budget. Your first computer may not be ideal, but if it gets you started on music production, then that is all that matters.

Terms

Avid Audio eXtension (AAX) The default plug-in format for the Avid Pro Tools DAW application. The plug-in format is only supported by Avid Pro Tools. The plug-in operates on MacOS and Windows operating systems.

Apple File System (APFS) The default file storage format for current MacOS computers.

Audio Units (AU) A plug-in format developed by Apple. The plug-in operates on supported DAW applications within the MacOS operating system.

Audio Stream Input/Output (ASIO) An audio device driver developed by Steinberg for Windows computers. The ASIO driver allows the DAW application to receive and send sound directly to a supported audio interface with minimal latency.

Bit The smallest unit of measure in a computer system.

Byte A file size consisting of eight bits.

Central Processing Unit (CPU) The main processor in a computer system that manages all connected devices and communicates with the operating system.

Core Audio An audio device driver developed by Apple for MacOS computers. The Core Audio driver allows the DAW application to receive and send sound directly to a supported audio interface with minimal latency.

Device Driver Computer code that allows a hardware device, such as an audio interface, to communicate and share information with a computer.

DisplayPort A digital display interface to connect computers to monitors. DisplayPort connections provide high-resolution graphics when used with support GPUs and monitors.

Extended File Allocation Table (ExFat) A file storage format that allows for large storage drive sizes. This storage format can be read and written to by MacOS and Windows computers.

Flash Storage A storage device that stores information on memory chips. Flash storage is non-volatile, so data is retained when the device is not receiving power.

Gigabyte A file size consisting of 1,024 megabytes.

Graphics Processing Unit (GPU) Commonly known as a graphics card. It is a computer hardware device responsible for rending graphics on a computer monitor.

Hard Disk Drive (HDD) Storage device consisting of rotating magnet platters.

High-Definition Multimedia Interface (HDMI) A digital display interface to connect computers to monitors. HDMI connections provide high-resolution graphics when used with support GPUs and monitors. HDMI can also carry audio information.

Kilobyte A file size consisting of 1,024 bytes.

Megabyte A file size consisting of 1,024 kilobytes.

Motherboard The main board on a computer. All computer components connect to the motherboard.

New Technology File System (NTFS) The default file storage format for current Windows computers. Can be read by MacOS computers.

Operating System (OS) A collection of programs that control the resources of a computer system written in a low-level language that speaks directly to the CPU, GPU, RAM, storage devices, and other components.

Personal Computer (PC) The term used by IBM to differentiate a home computer from a mainframe computer. Describes any computer running the Windows operating system.

Plug-In A software application that runs on top of an application adding features or functions. It requires a host application to function.

Random Access Memory (RAM) Physical memory used to temporarily store information. RAM is volatile and requires power to operate.

Solid State Drive (SSD) A storage device that uses flash memory to store data.

Terabyte A file size consisting of 1,024 gigabytes.

Thunderbolt A protocol that provides devices with direct connections to your motherboard for extremely fast transfer rates.

Universal Serial Bus (USB) A protocol that allows devices to be connected to a computer.

Virtual Studio Technology (VST) A plug-in format developed by Steinberg for MacOS and Windows computers.

Windows Driver Model (WDM) The default driver protocol for Windows computers allows hardware devices.

Chapter 5

Digital Audio Workstations

Digital Audio Workstations

The digital audio workstation (DAW) is the core component of your home studio. A DAW is a computer-based recording system that consists of three components: a computer, hardware, and software. DAWs are designed to behave like traditional recording equipment seen in a studio. The DAW performs the functions of a recording console, tape recorder, and effects processing in one package. These systems fall into three categories: self-contained systems, computer-based systems with dedicated digital signal processors (DSPs) and software, and computer-based systems with software.

The self-contained DAW is a dedicated computer-based system with a custom operating system designed solely for audio recording, editing, and processing. This system has a custom graphical user interface (GUI) that allows you to perform the functions of recording, editing, effects processing, and mixing. Many of these systems allowed you to deliver the final music project on a compact disc (CD) or an audio file. These systems support 8, 16, 24, or 32 audio tracks, which is the common configuration seen in recording studios with tape machines. Due to the closed nature of these systems, you have limited expansion options. If you purchase a 16-track system and later decide that you need 32 tracks, then you would need to purchase an entirely new system. The advantages of these systems are that they perform a single task which means all the processing power can be dedicated to working with audio. These systems allowed you to connect an external monitor, keyboard, and mouse so that you could operate the GUI with the same ease as a typical computer. Standalone systems were extremely popular in the 1990s with companies like Roland and Yamaha making affordable models. The increase in computing power in the last 20 years has reduced the need for standalone systems.

The computer-based system with dedicated DSP hardware and software is the second type of DAW. These systems start with a traditional computer running Windows or Mac operating systems. The computer handles the tasks of navigating the interface and drawing the screen. All the audio processing,

DOI: 10.4324/9781003345138-6

recording, editing, and mixing are handled by the dedicated DSP, which eases the amount of processing power required by the computer. The audio software decides which processes are handled by the DSP and not the computer. The result is a system that is stable, fast, and capable of handling a large number of audio tracks. These systems are ideal for situations where you are recording and mixing a large number of tracks, such as an orchestra performing film music. The most prevalent DSP system in the industry is the Avid Pro Tools HDX system. This system consists of the Pro Tools Ultimate software, an HDX PCIe DSP card, and Pro Tools audio hardware. While you can technically use other audio software on this system, such as Logic Pro or Cubase Pro, only the Pro Tools Ultimate software can access the DSP in the card. The Avid Pro Tools HDX system is ideal for large studios, but the price tag makes it cost-prohibitive for a home studio.

Computer-based systems that run all audio functions without dedicated DSP hardware are called native systems. The word "native" implies that all the audio processing is handled by the computer. This means that the computer must handle all operations required of the DAW software along with the normal tasks such as the operating system and other installed. Early native systems could only handle a limited number of tracks, and this was due to the speed of the central processing unit (CPU) or processor. With the advancement of multi-core and high processor speeds, native systems can handle many audio tracks. Native systems allow you to choose an audio interface of your choice and give you the flexibility to run any or many DAW applications. Native systems allow you to choose the DAW software that best suits your needs and provides you with the results you want. Your only limitation with a native system is the capabilities of your computer.

Now that we have a general idea of the different DAW types, let us look at the process of installing a DAW on your computer.

Installing and Configuring

DAW applications work best when you have a dedicated audio interface connected to your computer. If you do not have an audio interface, the DAW application can work with your computer's built-in sound card though there will be some limitations. In Chapter 7, Audio Hardware, we will discuss audio interfaces. For now, let us go through the general steps of installing a DAW on your computer.

Every DAW application has its own installation process. Some applications offer a single file that you must execute to begin installing. The installer will guide you through the installation of the application. Other applications will require you to install multiple items to enable all the features of the application. If you are unsure as to what features you should install, it is always best to use the default settings for the application. It is good practice to install all recommended features. Should you discover later that you do not need certain features, you can always remove them later. Since there is no single process that applies to all applications, it is best to read the installation instructions and follow the prompts during the install process. When the installation is complete, it is a good idea to reboot your computer. Rebooting your computer clears any temporary installer files and ensures that your system is ready to run the application.

After you install the application, you will need to run it for the first time. Some applications are very good at providing you with a "wizard" to guide you through the process of configuring your audio interface and other peripherals attached to your computer. You will discover that others are not as friendly and will require you to navigate a series of menus to manually configure your audio interface as well as your inputs and outputs. Once again, follow any included instructions to guide you through this process.

If your DAW application includes a demo project or session, then it is advantageous to open that project after you configure your audio interface and peripherals. Manufacturers will often provide you with a demo session that uses most, or all the features and plug-ins included with the application. This is a good way to test your installation and make sure you installed all the features for the application. The demo session also provides you with the opportunity to check your audio output and make sure that you can hear sound from the application.

Now that your DAW application is installed, we can begin to talk about the features and functions that are common to all DAW applications.

Features

You may never have visited a recording studio or seen one in action; therefore, you may be unsure as to what tasks should a DAW be able to perform. For audio software to be classified as a DAW application, it must provide a basic feature set that mimics what you can do in a recording studio. First, you need the ability to record and store audio for later reference. After you record the audio, you need to be able to edit the recordings and then make additional recordings to supplement the initial recording. For example, you might first record drums for your song. Once you edit the drums so that the performance sounds the way you like, you may ask the bassist and guitarist to record their parts against the drum tracks. After you finish all your recordings and edits, you then need to mix all the sources so that they complement each other and sound well together. During the mixing process, you may wish to add effects such as reverberation and delay to the instruments to create ambience. You may want to add or remove certain frequencies from the sounds to make them easier to hear. This is known as effects processing. When you finish your mix, you need to create a stereo version of your song that can be played and shared on different media devices. This final stage is sometimes called "bouncing," which is the process of taking several tracks and combining them into a two-track stereo version.

To accomplish all these tasks, the software must offer other functionality. The software must provide transport controls such as stop, play, fast-forward, rewind, and record. There should be a main clock that is constantly visible and can measure time in different formats as well as measure numbers. The application should support a variety of tracks: audio, instrument, MIDI, auxiliary, and master. The mixing interface should resemble a recording console allowing you to see multiple faders and control the volume of each track and knobs to control the panning, or stereo placement of each track. The application should provide you with a variety of audio effects plug-ins so that you can shape and alter each track as you wish. The editing process should be intuitive and logical so that you can edit and manipulate single or multiple tracks at the same time. Editing functions should include copy, paste, cut, delete, and other commands. The ability to record single or multiple tracks at the same time is also important. Finally, the application should give you the flexibility to route audio to any source or destination whether internally or externally through your audio interface.

All DAW applications on the market offer these essential features. Each DAW application may execute these functions differently. However, once you understand how to operate one DAW, the ability to learn a second, or even third application becomes easier. The issue is not whether the application can perform the tasks you want, but rather how they perform that task. Some applications will automate certain tasks for you while others require you to manually configure and set up each stage of the process. How and where items are displayed also varies between applications. One application may place the transport at the top of your window, while another places it at the bottom, and others give you the option of where you want the transport placed. These are subtle distinctions, but I do want you to be aware that every DAW you encounter will look and feel different.

You may be wondering which is the best DAW on the market. Maybe your thinking is more practical, and you want to know which DAW gives you the best value for the money. Fortunately, there is no single best application to choose from. Each application has its own strengths and perhaps weaknesses depending on your personal preferences. What you see as a shortcoming in a particular application may not be perceived the same way by another user. Fortunately, many manufacturers offer time-limited demo versions of their software for you to try with no financial commitment. This is the best way to experiment with different applications and find the one that best meets your needs.

Now that we have an idea of what a DAW application should offer us, let us explore these features in greater detail.

Transport

The transport provides the controls to navigate your project. Common controls are rewind, fast-forward, stop, play, and record. Additional controls may include previous marker, next marker, start of project, and end of project. These controls resemble those seen on tape decks in recording studios. The transport also offers additional functions such as tempo and meter indications, and the ability to enable the metronome click. The transport is connected to the main clock and timeline of your project (Figure 5.1).

Main Clock

The main clock or main display will tell you what measure your playback line or cursor is currently located. The display should give you the option to view measures or a variety of time formats. Basic time formats could include hours, minutes, and seconds while more advanced time formats with support timecode and video frames. More flexible displays allow you to see the measure number and the corresponding time within the same window. The main display will allow you to manually type in a measure number or time to advance your playback cursor or line to that point. Some displays will even indicate the number of measures you have selected for an edit or a range selection. Your display should be easy to read and located either at the top or bottom of the screen. Some applications will give you the ability to float the display so that you can place it anywhere you like within your window (Figure 5.2).

Timeline

The timeline runs along the top of your main window and the current location of your playback cursor or line. The timeline provides you with the

Figure 5.1 Steinberg Cubase transport (Courtesy of Steinberg Media Technologies GmbH, a division of Yamaha Corporation, www.steinberg.net)

Figure 5.2 Steinberg Cubase main display (Courtesy of Steinberg Media Technologies GmbH, a division of Yamaha Corporation, www.steinberg.net)

ability to click on a specific location and begin playback or recording from that point. The timeline also allows you to select a range and playback the selected range. The timeline can display measure numbers or minutes and seconds. Some DAWs allow you to select timecodes, samples, and video frames as your linear reference. Other DAWs allow you to display multiple timelines so that you can see your current bar and current time.

Tempo, meter, and key signatures are often included as additional timelines. These are all displayed linearly across your screen and provide you with detail and control. This can be useful if your music changes tempo, meter, or key at different points. Do you really need all these timelines? If you are a film composer, you might add markers to mark out different scenes in the film you are working on. You may also need to see the timecode so that you can reference specific locations from a director. The type of music or production you do determines your needs; many of us are quite happy with just measure numbers or minutes and seconds (Figure 5.3).

Tracks

The DAW should support several different track types within your project or session. Audio tracks and MIDI tracks are the most common. Audio tracks are needed when recording external sounds such as a piano, guitar, drums, or voice. MIDI tracks are useful for recording MIDI data from a hardware synthesizer. Virtual instrument plug-ins included with your DAW require instrument tracks to host the audio and MIDI data. Auxiliary or Bus tracks are useful for routing audio from multiple locations to a single track or hosting an effect plug-in that can be shared with other tracks. Folder tracks are useful for grouping tracks for organization and editing.

If your project has tempo changes, then a dedicated tempo track allows you to specify where tempo changes occur. Earlier I mentioned that tempo changes can be mapped to the timeline. This is a difference between DAWs; some allow you to track tempo in the timeline, while others use a dedicated track for tempo. Both offer the same functionality; it is just a preference determined by the manufacturer. Time and key signatures are also options for tracks for the DAWs that do not show these items in the timeline.

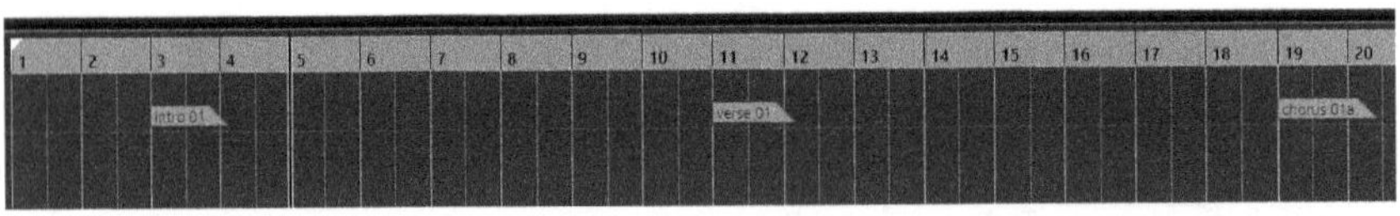

Figure 5.3 Steinberg Cubase timeline (Courtesy of Steinberg Media Technologies GmbH, a division of Yamaha Corporation, www.steinberg.net)

The number of tracks for each type is determined by the manufacturer. Many manufacturers offer different versions of their DAW such as intro or entry, standard, and professional. The intro version may only offer eight or 16 for each track type, the standard 32 or 64, and the professional 128 or more (Figure 5.4).

Mixing

This mixing environment in a DAW is reminiscent of a traditional mixing console found in a large studio. There are individual channel strips for each track. Each strip contains a series of controls. The most common is a fader to adjust the volume of the track and a pan knob to determine the stereo placement of the track. The strip has a section often called Inserts. The Inserts allow you to add an audio effect plug-in such as reverberation, chorus, compression, or equalization to the track. There will usually be several Insert slots so that you can add multiple effects to a track. Sends on the strips allow you to split the signal so that you can send it to another location, such as an auxiliary effects track and still have control of the original sound. The channel strip also features a Mute button to temporarily silence the track and

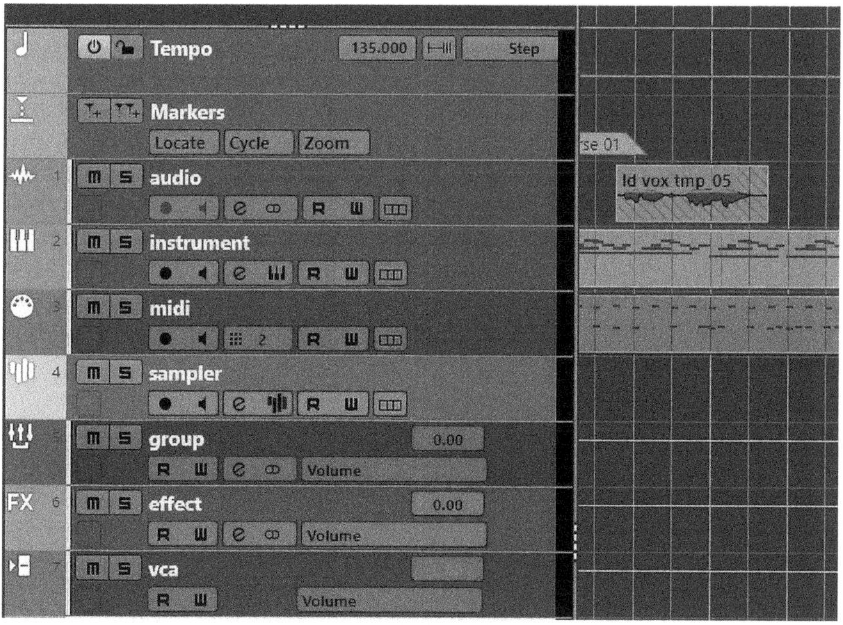

Figure 5.4 Steinberg Cubase track types in the timeline (Courtesy of Steinberg Media Technologies GmbH, a division of Yamaha Corporation, www.steinberg.net)

remove it from the mix. The Solo button allows you to isolate the track from the others temporarily so that you can focus on listening to the single track (Figure 5.5).

Each channel strip has input and output routing controls to determine which audio input on your interface the sound is entering from, and which output the track is routing sound. This feature gives you the ability to route audio to and from any location. You can route audio to a physical output on your audio interface or a virtual one within the DAW application. You also can route any of your audio inputs to any or multiple audio tracks and have the ability to record in mono, stereo, or if your interface supports it, other multichannel formats. The ability to route audio should be flexible enough for you to complete any task that you wish.

Recording

Each strip also has controls useful when recording audio or MIDI to a track. A record or arm button is present on each channel strip. The record or arm button is a red circle, which is the industry standard for recording. A monitoring button allows you to hear the input signal coming into the track without needing to arm the track. This button determines whether you are listening to the information already recorded on the track or the input source. A DAW will allow you to record several tracks at once, meaning you can arm multiple tracks and record unique information on each one if your audio interface has multiple inputs.

One of the main advantages of working with a DAW and the digital environment is the ability to record multiple versions of something on a track without erasing a previous version. If you are recording a vocalist for a song, you can have them record their part several times. Each version or take is stored as a separate lane on a single track. When you have finished recording the singer, you can view the lanes and listen to each take individually. From there you can determine which sections are the best and comp, or compose, the best version of the vocal track. Editing is another advantage of working with a DAW (Figure 5.6).

Audio Editing

Audio editing is at the heart of every DAW application. Basic editing functions such as cut, paste, and undo should be easy to perform and use common keyboard shortcuts that you are already familiar with. Advanced editing features include the ability to change the timing and pitch of audio. You can select a portion of the audio and edit it or place it at another location. I already mentioned the ability to record multiple versions or takes of a performance and then select the best portions of each take to create a complete version of that performance. Some editing requires more precision, so the

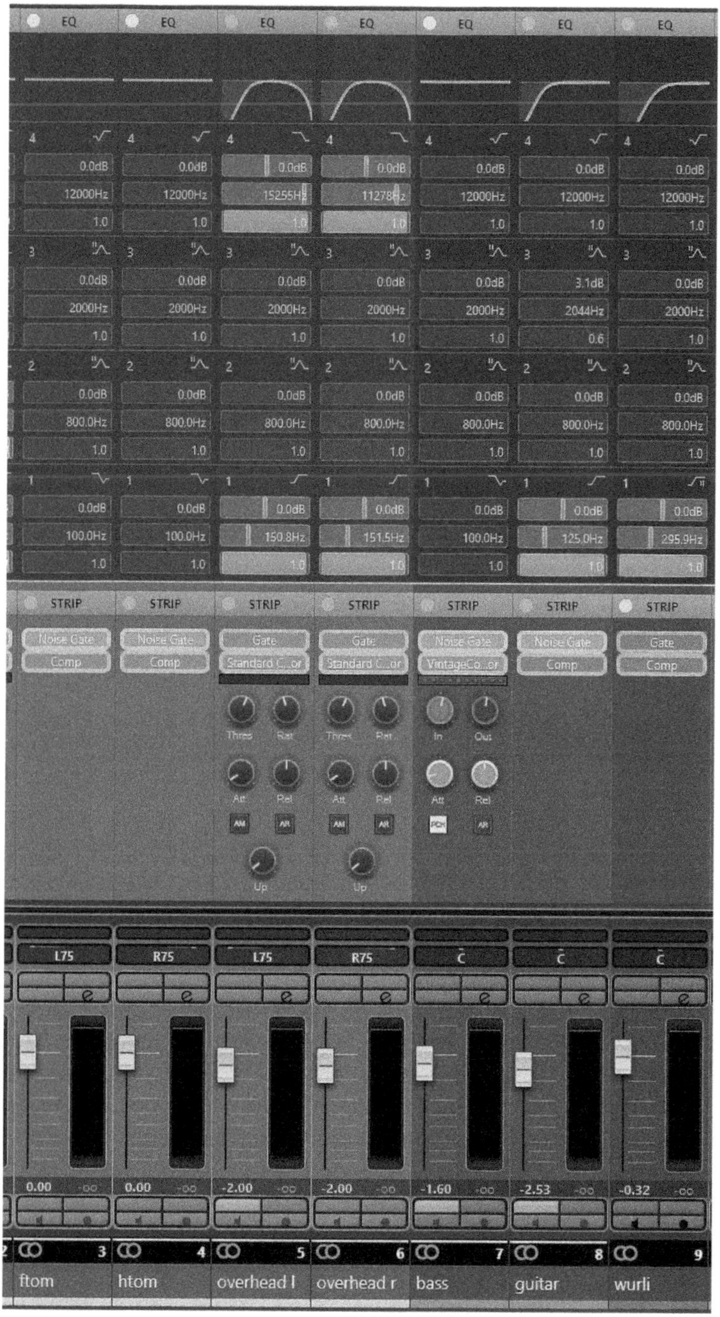

Figure 5.5 Steinberg Cubase Mix console (Courtesy of Steinberg Media Technologies GmbH, a division of Yamaha Corporation, www.steinberg.net)

Figure 5.6 Steinberg Cubase channel strip (Courtesy of Steinberg Media Technologies GmbH, a division of Yamaha Corporation, www.steinberg.net)

ability to zoom in to a desired level so that you can see the information that you need to see to perform your edits is crucial. Once you finish your editing, you can zoom out so that you can see all your information on a single screen. If an edit does not go as planned, you can undo the edit and return to a previous state.

Nondestructive editing is perhaps the most significant and important feature in a DAW application. Nondestructive editing means that regardless of what you do to the audio represented on the track, the original audio file on your hard drive remains unchanged. Since the original audio file is unchanged you can perform multiple tasks and easily undo those tasks to return to a previous editing state. Nondestructive editing prevents the creation of multiple audio files that could fill up your hard drive and affect your computer's performance.

The principle behind nondestructive editing is simple. The audio that you see on any given track is a representation of the file that is stored on your hard drive. When you select and press the delete key, that portion of the audio is removed from the visual representation on the track. The original audio file is unaffected. The edit you made tells the DAW to not play that portion of the audio. If you copy a section of audio and place it later in the track, the DAW will playback the copied audio from its original location. Nondestructive editing is wonderful because you never have to worry about deleting or removing the actual audio file from your hard drive. You can always return to your previous state.

There are instances, however, where you will want to create a new audio file to represent all the edits that you have made. For example, when you finish comping the vocalist I mentioned earlier, you may want to create a single audio file that consists of all the edits you made. The DAW will allow you to bounce or render the selected track as a new audio file. This new audio file will be added to your project folder. You can archive the comped track should you need to go back and correct some edits.

Please keep in mind that the application still can permanently delete the audio file from your hard drive. Rest assured, however, that most applications will present you with a series of reminders that you are about to actively destroy the audio file. DAW applications with advanced features give you the ability to create a copy of your project that only contains the audio files that are being used in the project. This process is often called consolidation. Applications with advanced features provide you with full file management control over the files on your hard drive and where they are stored. It is always a good idea to make a backup of your project and audio files in case something happens to your computer (Figure 5.7).

MIDI Editing

The ability to edit MIDI data is essential to a DAW. When you work with virtual instruments and hardware synthesizers, you will be working with MIDI data. The DAW needs to perform simple and complex MIDI tasks. The DAW needs to assign MIDI channels to individual tracks and assign individual instruments to each track. Note entry is accomplished using real-time entry where the musician plays the instrument while following a click track.

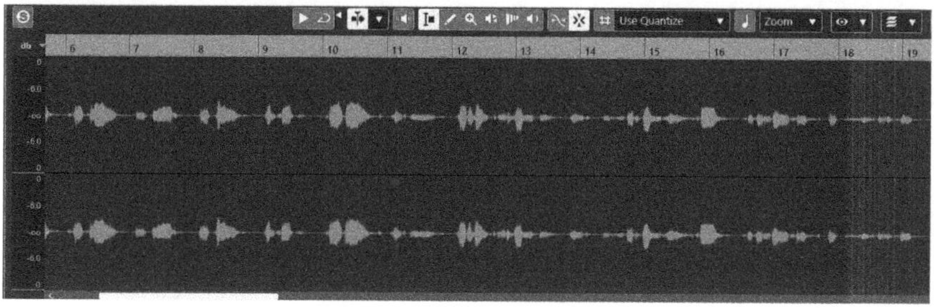

Figure 5.7 Steinberg Cubase audio editor (Courtesy of Steinberg Media Technologies GmbH, a division of Yamaha Corporation, www.steinberg.net)

Another option is called step-entry, which allows you to enter each note individually specifying the pitch, timing, and duration. Step entry is extremely useful when creating complex MIDI parts.

The ability to edit MIDI data is also important. The DAW must provide you with the ability to manipulate MIDI data using copy, cut, and paste methods. The DAW must allow you to specify the key, tempo, and meter of the project and have the freedom to change any of those within the same composition. The ability to transpose notes up or down by intervals is helpful when looking for the best range or key for a musical passage. You also needed the ability to enter and edit MIDI expression information such as velocity and sustain.

The timing of MIDI information is measured in ticks. A tick is a subdivision of a music quarter note. The most common subdivisions of a quarter note are 480 or 960 ticks. A DAW using 480 ticks means that a quarter note is divided into 480 parts. The smaller the note value the fewer number of ticks. An eighth note would be 240 ticks and a sixteenth note 120 ticks. Ticks control the timing of MIDI tracks, but the subdivisions allow performances to be captured more accurately; the greater the number of ticks the more accurately the performance is recorded.

The ability to correct the timing of a performance and align it to a musical grid is extremely popular, especially when creating drum sequences. This process is called quantizing. When you quantize MIDI information, you are aligning MIDI information, such as notes, to the grid in ticks. You can quantize to a specific note value, such as a sixteenth note. This means that all MIDI data will be aligned to the closest sixteenth note in either direction. When MIDI data is quantized exactly, drum and music passages can feel rigid and mechanical. To counter this effect, MIDI data can be "humanized" by randomly altering the timing by small increments. These small alterations reduce the mechanical feel of MIDI passages (Figure 5.8).

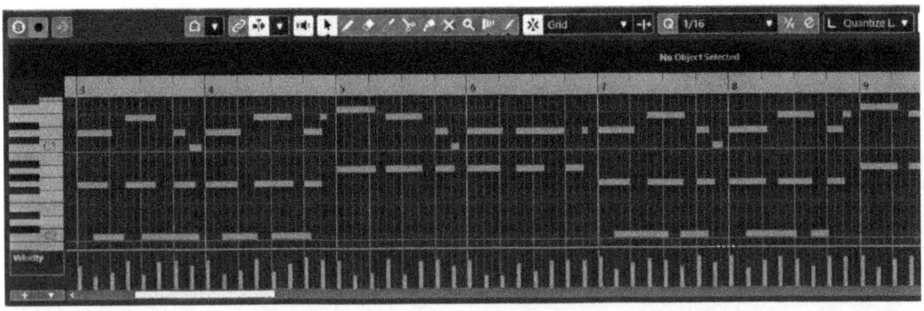

Figure 5.8 Steinberg Cubase MIDI editor (Courtesy of Steinberg Media Technologies GmbH, a division of Yamaha Corporation, www.steinberg.net)

Effects Processing

Earlier, I mentioned that a channel strip has inserts where we can add audio effect plug-ins to process the audio on a given track. DAW applications include a variety of audio effects plug-ins. The number, variety, and quality of these effects vary by manufacturer, but the selection is often diverse enough for you to process your audio without having to purchase additional effect plug-ins. I recommend familiarizing yourself with the included effect plug-ins before spending money on third-party effects. You might be surprised by the choices included with your application. We will cover audio effects in the next chapter.

Virtual Instruments

DAWs include a selection of virtual instruments for use on instrument tracks. Virtual instruments are software representations of synthesizers, drum machines, and other electronic instruments. Virtual instruments can represent vintage physical synthesizers or be original creations based on different synthesis techniques. DAWs may also include software samplers. A software sampler plays back recordings of real instruments such as pianos, guitars, basses, and orchestral instruments. Sample libraries vary in size and quality.

Virtual instruments allow you to add additional sounds to your recordings that you may not be able to access. You might be working on a song that needs a piano. The software sampler in your DAW can provide you with a piano library to work with. The same song could use a drum track for a live drummer to reference. The virtual drum machine included with your DAW can help you create one.

Virtual instruments are helpful for composers and songwriters. We will look at synthesizers, samplers, and virtual instruments more closely in a later chapter.

License Management

DAW applications and plug-ins are intellectual property. You are not allowed to own intellectual property. When you purchase an application or plug-in, you are purchasing a license to use the product. The conditions of the license vary between manufacturers and are explained in the End-User License Agreement (EULA). This agreement specifies the conditions under which you may use the product. The EULA will tell you the number of computer installations allowed. It will also inform you whether you can sell the product to another user. The EULA may even notify you that the license will expire after a certain period. When you install the product on your computer, you agree to not copy or share the software with another user. You also agree not to modify or reverse engineer the software code. The EULA is a legal document.

This might seem excessive, but software piracy is a concern for all application and DAW developers. Companies spend money to research, develop, market, and sell software. Software piracy means that the company is not making money from the use of the product. With little effort, you may find "cracked" versions of the application or plug-ins you want to use. When someone finds a way to alter the copy protection code on an application or plug-in, then that product has been cracked. Many users will justify using cracked products by saying that they will pay for it when they can afford it or that since they are not making money from the product, they do not need to pay for it. Using cracked software is illegal and harms the music product industry. If you wish to work in the music industry as a producer, artist, or recording engineer, then you should be part of the industry and purchase the products you use.

Copy Protection

To prevent users from making copies of software products or sharing them with others, developers add copy protection to the products. Copy protection is a layer of security that restricts the use of the product to a specific computer. Copy protection keeps track of the software location and can disable access to it should something change. There are different types of copy protection implemented in applications and plug-ins.

Serial Number

The serial number method is the easiest and most straightforward method of copy protection. When you first run the software, you will be asked to enter a serial number. If the serial number matches what the product is expecting, then the product will work. The serial number method is the least secure because serial numbers can be easily shared. On the user end, serial numbers can be lost. Serial number protection can be enhanced by requiring an internet connection to verify the serial number.

Challenge/Response

The challenge/response method begins with a serial number. The software will contact a database server to verify the serial number and then either send an authorization code or automatically authorize the software. The advantage of this method is that the database can track the number of authorizations issued. Most challenge/response systems will allow you to de-authorize a product so that you can install it on another computer. If you exceed the maximum number of authorizations, you need to contact the manufacturer to gain more authorizations. This can happen if you have a computer crash and are unable to de-authorize the software. The challenge/response method stores the license on the computer.

Hardware Licenser

A hardware licenser is a physical Universal Serial Bus (USB) device that stores the license. If the device is connected to your computer, you can use the software. The USB device can be placed on any computer, even one that you do not own, and the software will work. This is convenient if you are working in another studio and need to use your plug-ins on that machine. You can install the plug-ins on the computer and use them without violating the EULA. Once you remove the USB device, the plug-ins will not work. The danger of using a USB device is that if you lose the device, you lose access to the software. Some manufacturers allow you to use any USB thumb drive to store the license. Others require a specialized device, such as an iLok.

ILok

The iLok license manager is probably the most popular method for license management. The iLok system stores the license on an iLok, which is a USB stick, resembling a thumb drive. The iLok device can only store licenses managed by iLok and requires the iLok License Manager software to authorize and de-authorize software. The iLok License Manager allows you to authorize multiple iLoks associated with your account. The company even allows you to purchase insurance so that in the event you lose your iLok device, you can retrieve your license without delay.

The iLok system allows manufacturers to choose two alternate methods of storing the license on the iLok device. The first method is to store the license directly on the computer. Even though you do not need the iLok device, you still must create an account with iLok and install the iLok License Manager. When you want to use the license on another computer, you must de-authorize the license from the first computer, and then authorize it on the second. You must repeat the process if you wish to restore the license to the first computer.

The other alternative is to use the iLok Cloud. The iLok Cloud stores the license on a server on the internet. When you launch the software, the iLok License Manager connects to the iLok Cloud. When it locates the license, it authorizes the connected computer, and you can use the software. When you have finished using the software, the license is returned to the iLok Cloud. The method allows you to use the software on multiple computers, although not at the same time. The iLok Cloud requires a constant internet connection to function. If you are working remotely without an internet connection, you cannot use the iLok Cloud.

Your iLok options vary from manufacturer. Many offer all three options for you, but not all. Some manufacturers grant you between one and three authorizations when you purchase the software. Multiple authorizations mean that you can license your laptop and desktop and do not need to use the iLok Cloud or USB device. The advantage of the iLok device is that if your computer crashes, the licenses are stored on the device. When you restore a computer or purchase a new one, you install the software, connect the device, and you are ready to go.

Copy protection is a reality of the music software industry. It may feel inconvenient and unnecessary, but it does benefit everyone. Purchasing software helps the music industry and ensures that these companies continue to develop products for us to use. If someone offers you cracked software, kindly reject the offer. Support the industry that you work in.

Summary

In this chapter, we looked at the DAW and how this tool is the foundation of your home studio. The DAW replicates all the functions of a multichannel recording console and a multichannel recording device. Each DAW has features that differentiate it from another brand, but all of them share similar functions and controls. The best DAW is the one that best fits your workflow. As your skills develop, you may find yourself using different DAWs for different tasks. This is not uncommon as one product might be better suited for a particular style of music or workflow than the other.

Activity 5

Audio Editing

In this activity, we will use Audacity to edit an audio file and save the edits in different audio file formats.

Please go to the companion website and download the WAV file for Activity 5. This audio file has been generously provided by the songwriter and performer of the song, Richard Parlee. Please understand that this is copyrighted material and cannot be used for commercial purposes.

Before launching Audacity, create a folder on your computer, either on the desktop or in your documents folder. Name the folder *audio editing* and place the downloaded audio file in that folder. Now we can begin the activity.

1 Launch Audacity

   a From the menu, go to File – Open
   b Go to the location of the folder you created
   c Select the file and click Open
   d Use the magnifying glass with the plus sign in it to zoom in as needed

2 Go to File – Save Project – Save Project As...

   a You will get a warning that a project does not create audio files, click OK
   b Select the project in the folder you created earlier
   c Click Save

3 Our first edit will cut out the beginning instrumental of the song

   a Place your cursor at around 29 seconds on the timeline
   b Place the cursor above or below the center line
   c The mouse pointer will turn into a vertical line
   d Click, hold, and drag to the left until you reach the beginning of the file
   e The selection will be highlighted
   f Press Delete on your computer keyboard to remove the audio

4 Rewind to the start of the file and press the play button

   a The audio file starts abruptly

5 We will add a fade in at the beginning so that the beginning does not sound abrupt

   a Zoom in more so that you can clearly see the beginning of the file
   b Place the cursor just before the singing starts
   c Click, hold, and drag to the left until you reach the beginning of the file
   d The selection will be highlighted
   e Go to Effect in the menu bar at the top and select Fade In

6 Rewind to the beginning of the file and press play

   a The music will now fade in

7 We will now shorten the length of the song

   a Place the cursor at around 2 minutes and 39 seconds
   b Click, hold, and drag to the right until you reach the end of the file
   c Press Delete on your computer keyboard to remove the audio

8 Now we can add a fade out so the song gradually fades away

   a Place the cursor around 2 minutes at 31 seconds
   b Click, hold, and drag to the right until you reach the end of the file
   c The selection will be highlighted
   d Go to Effect in the menu bar at the top and select Fade Out
   e Listen to the file and notice how the song fades out gradually

9. Go to File – Save Project – Save Project
10. Let us take a moment to see how the files look in the project folder we created
    - a In Windows Explorer or Mac Finder, navigate to the place where you stored your Audacity project
    - b Notice how the WAV file is still intact and unchanged even though we made two edits on the file
    - c This is because the Audacity project file stores all the edit information
    - d Audacity allows us to perform non-destructive edits
    - e The original audio file is unchanged
11. If we want to create a new WAV file of our edits, we need to export to a new WAV file
    - a In the menu, go to File – Export – Export as WAV
    - b Create a unique filename for your edited file
    - c Save the exported file in the same folder as your Audacity project
    - d Under Encoding at the bottom, notice that it says Signed 16-bit PCM
        - i This just tells us that the file will remain at 16-bit, which is how we added it to the project
    - e We have the option to enter metadata
        - i This tells us that the file will be saved as a broadcast WAV file
        - ii We can leave this blank for now
    - f Click OK
    - g The new file is saved to our project folder
12. If we compare the file size of the edited file to the original, you will see that it is smaller
    - a This is because the new file is shorter in length than the original
13. We can use Audacity to export to MP3 as well
    - a From the menu, go to File – Export – Export as MP3
    - b The MP3 export window is similar to the WAV export window
    - c The file extension is now MP3 and we can choose our MP3 streaming rate
14. We can use the default settings, but let us customize
    - a Select Constant for the Bit Rate Mode
    - b Choose 256 kbps for the Quality

c Choose Stereo for the Channel Mode
d These settings provide the best compromise for MP3 files in that it gives us a good-sounding file for the size
e Click Save

15 This time we will fill in the metadata

f Artist Name: Richard Parlee
g Track Title: Feel This Way
h Album Title: Out of Slumber
i Track Number: 10
j Year: 2020
k Genre: Folk

16 Click Save

17 The MP3 file is created

Double-clicking on the new MP3 file will open the default media player on your computer. Your media player should display the metadata for the file.

As you can see, Audacity is a very useful audio editor. You can perform a variety of audio editing tasks. This application is handy and worth keeping around.

This concludes this activity. If you want, you can save the project or quit without saving.

Terms

Copy Protection A method that prevents an application from being copied to another computer or have its licensing altered.

Destructive Editing When the original audio file is altered with each edit performed.

Digital Signal Processor (DSP) A device that processes audio information separately from the computer CPU.

Native System A digital audio workstation that runs entirely on a computer without any DSP.

Nondestructive Editing When the original audio file is note altered with each edit performed.

Quantize The process of aligning sound to a timing grid

Tick A subdivision of a quarter note used by DAW applications. Most DAW applications default to 480 or 960 ticks per quarter note.

Virtual Instrument A software representation of synthesizers, drum machines, and other electronic instruments.

Chapter 6

Audio Effects

Effects Processing

This chapter provides a basic overview of effects processing. The goal is to provide you with a basic understanding of effects and how they are commonly used. We will cover the common controls found on effects and their parameters. The companion website provides you with audio examples so you can hear the difference between the effects. The companion website provides you with additional resources if you wish to explore effects processing in greater detail.

Audio effects can be broken down into three effect families: dynamics, frequency, and time. Dynamics effects help us control the difference between the loudest and quietest sound a track produces. Frequency effects can boost or attenuate specific frequencies from a track. Finally, time effects change our perception of sound so that it can sound far away or in a large open space. Let us look at each family and explore the various effects within these families.

Dynamics

Dynamics processors are effects that allow you to control or alter the dynamic range of an audio signal. Dynamic range is defined as the difference between the quietest and loudest sound that an audio signal can produce. A dynamics processor can give you the ability to manage a recording that varies in volume. For example, a bass guitar recording might be consistent throughout the verses, but during the chorus, some of the notes are louder than the others. A dynamics processor can control those loud notes by lowering the volume of any note that is louder than a specified threshold. On another track, you have an electric guitar that hums whenever it is not playing. You can use a dynamics processor to prevent any sound from the guitar from being heard unless it is louder than a specified threshold. A dynamics processor is useful in managing a track when the fader does not provide enough control. If the overall guitar track is too loud, then the easiest and most practical solution is to lower the fader for that track so that the guitar does not sound as loud.

DOI: 10.4324/9781003345138-7

If the guitar track is loud in some sections and quiet in other sections, then a dynamics processor can help you find a balance between both extremes.

The simplest dynamics processor is the gate. A gate only allows sound to pass if it is above a certain volume level. If we go back to our guitar track with a hum, we can use the gate to prevent the hum from being heard whenever the guitar is not playing. The gate has three controls: threshold, attack, and release. The threshold is the volume level that triggers the gate to open and close. When the signal on the track goes above the threshold, the gate opens. When the signal falls below the threshold, the gate closes. The attack determines how quickly the gate opens once the signal exceeds the threshold level. The release determines how quickly the gate closes after the signal falls below the threshold. In the case of our guitar track that has a hum, we would set the threshold at a level such that the hum is below the threshold. We would set a fast attack so that the gate opens quickly once the guitar starts playing at a level above the threshold. The release time varies depending on how quickly we want to gate the close. If the guitar is playing short notes, then a fast release time is better whereas a slower one works best for long notes. Noise gates are extremely useful for preventing unwanted sound from being heard on the track (Figure 6.1).

Figure 6.1 Steinberg Cubase gate plug-in (Courtesy of Steinberg Media Technologies GmbH, a division of Yamaha Corporation, www.steinberg.net)

A compressor is an effect that reduces the dynamic range of an incoming signal. Many users believe that a compressor makes an audio signal quieter. A better way of understanding a compressor is that it prevents the sound from being too loud. A compressor accomplishes this by reducing the dynamic range of an audio signal; it makes the loud signals less loud. A compressor typically has four settings, three are shared with the gate: threshold, attack, release, and ratio. Like to gate, the threshold on a compressor determines the level above which the compressor will engage. Signals that are above the threshold level will cause the compressor to activate. Signals below the threshold are unaffected. The attack time will determine how quickly the compressor reacts once the signal passes a threshold. Sounds from a drum set are immediately loud, so the attack time for those sounds could be fast. However, if the attack time affects the initial transient or attack of the drum set, then a slower attack is needed. The release time determines how long the compressor affects the dynamics of a sound after the level falls below the threshold. Sounds that die out quickly, like a snare drum might require a fast release time. A cymbal will continue to ring after it is struck so a longer release time might be better. Ultimately, you must listen to the sound and if the attack or release adversely affects the sound, then adjust the settings.

The ratio control on a compressor determines how much the incoming signal is reduced once it crosses the threshold. Unlike a linear setting, where the level is reduced by the same amount, for example, 2 dB, regardless of how far above the threshold the incoming signal is, the ratio allows for a gradual reduction. Signals that exceed the threshold by a small amount are reduced less than those that exceed the threshold by a larger amount. Remember that in digital audio, the loudest any signal can be is 0 dB. Signal level is measured with negative numbers so -20 dB is louder than -40 dB. Let us set the threshold for our compressor at -20 dB. With a linear setting of 2 dB, a signal that is -18 dB would be lowered to -20 dB, whereas a signal at -10 dB would be lowered to -12 dB. If we use a ratio of 2:1, then signals 2 dB above the threshold will only increase by 1 dB above the threshold. Using ratios, the -18 dB signal would be lowered to -19 dB because the signal is 2 dB above the threshold and the ratio will only allow the signal to be 1 dB above the threshold. A -10 dB signal would be lowered to -15 dB because the signal is 10 dB above the threshold resulting in a 5 dB increase above the threshold. The amount a signal is lowered by the compressor depends on how many decibels above the threshold the incoming signal is. The signals are not being lowered by the same amount but rather lowered proportionally to how loud they are (Figure 6.2).

The last dynamics processor we will examine is the limiter. A limiter is an extreme version of a compressor. A limiter does exactly what its name states: it limits the loudness of an audio signal being output. The limiter acts like

Figure 6.2 Steinberg Cubase compressor plug-in (Courtesy of Steinberg Media Technologies GmbH, a division of Yamaha Corporation, www.steinberg.net)

a ceiling and prevents all sound from passing that ceiling. Many engineers abuse the limiter because it allows them to increase the volume of their tracks without distorting the signal on the output. The effect of a limiter is dramatic and when used improperly can completely remove or restrict the dynamic range of the sound being fed through it. The limiter shares the same settings as a compressor except that the ratio is set to 100:1 or higher.

If you set a limiter's threshold to -20 dB, then regardless of how loud the signal is coming into the limiter it will never be louder than -20 dB. Regardless of whether the incoming is at -10 dB or -5 dB, the resulting output will always be -20 dB. Limiters are very useful in live performances because they can be set to prevent a performance from exceeding a specific volume level. They can also help protect monitors from receiving too much signal and distorting. Depending on the genre of music you are working with, the limiter can be a useful tool (Figure 6.3).

Figure 6.3 Steinberg Cubase limiter plug-in (Courtesy of Steinberg Media Technologies GmbH, a division of Yamaha Corporation, www.steinberg.net)

Frequency

Equalizers affect frequencies in sound. They can control the amount of energy of a specified frequency or a range of frequencies. Equalizers can make frequencies louder (boost), quieter (attenuate), or remove them entirely. It is critical to understand that equalizers cannot add frequencies that do not exist in the sound. For example, if a kick drum or bass guitar is lacking low frequencies in the recording, you cannot use an equalizer to add the missing low frequencies. Equalizers can only affect the sound that exists. Equalizers fall into three categories: filters, graphic, and parametric.

Filters are the simplest because they affect a range of frequencies and can either boost, attenuate, or remove frequencies in the given range. Shelving filters boost or attenuate all signals above or below a selected frequency. They are called shelves because they affect all frequencies in the selected range equally. These are basic tone controls and can often be found on consumer stereo audio systems such as bass and treble. A low shelving filter allows you to specify a low frequency, around 120 Hz and below, and either boost or attenuate those frequencies. A high shelving filter affects high frequencies, around 5,000 Hz and above, and boosts or attenuates that range. A shelving filter cannot remove all the frequencies in the specified range. If you want to remove frequencies above or below a specific frequency, then you must use either a high-pass or a low-pass filter (Figure 6.4).

A high-pass filter allows frequencies above a specified frequency, called the cutoff frequency, to pass and be heard. High-pass filters are sometimes called low-cut filters, which can be a little confusing, but they describe the same process. If I set a high-pass filter with a cutoff frequency at 5,000 Hz, then all frequencies above 5,000 Hz will be allowed to pass, and the frequencies below will be attenuated. Both terms describe the same result. A low-pass filter, or high-cut filter, allows frequencies below the cutoff frequency to be heard. A low-pass filter set at 120 Hz allows frequencies below 120 Hz to pass while frequencies above will be attenuated (Figure 6.5).

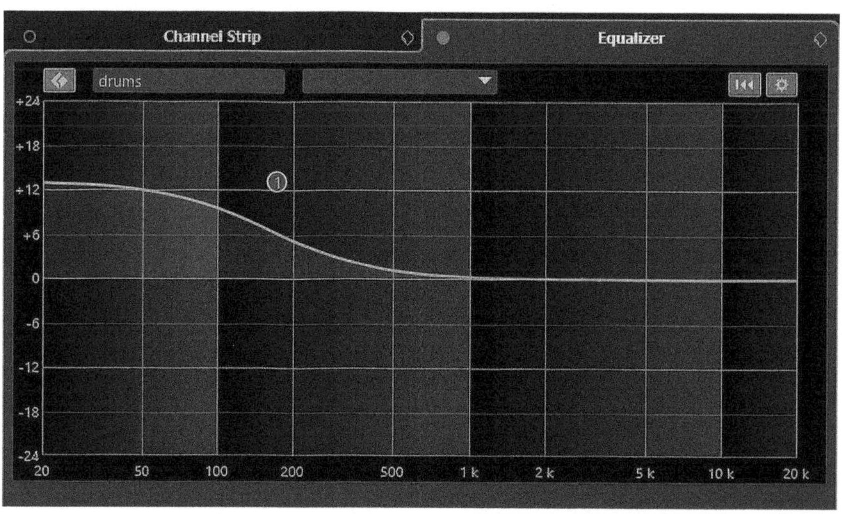

Figure 6.4 Steinberg Cubase shelving filter plug-in (Courtesy of Steinberg Media Technologies GmbH, a division of Yamaha Corporation, www.steinberg.net)

When you compare the shelving and pass filters in the images given in Figure 6.5, notice how they affect frequencies differently. The shelving filter name is appropriate because all the frequencies above or below are treated equally. Notice that the shelf never reaches the bottom, meaning the frequencies are lowered and never removed. The pass (or cut if you prefer) clearly extends to the bottom, removing frequencies above or below the cutoff frequency.

Now look at the line at the cutoff frequency. Notice that the line is not a straight vertical line, but rather an angled line. This angled line is called the slope. The slope is the angle of the ramp, which can be changed to be gradual or steep. The slope makes the boost or attenuation of the frequencies sound more natural and less abrupt. The angle of the slope is determined by how many decibels each octave above the cutoff frequency is lowered by. The octave of a given frequency is determined by doubling the frequency amount. If your cutoff is set at 110 Hz then the next octave up is 220 Hz, followed by 440 Hz, and so on. A slope of 6 dB means that the volume level at 200 Hz will be 6 dB lower than that at 100 Hz and the volume at 400 Hz will be 12 dB lower than that at 100 Hz. The higher the dB per octave, the steeper the ramp becomes. A 12 dB slope is said to be the most natural-sounding slope to our ears (Figure 6.6).

Graphic equalizers use regularly spaced, fixed-frequency filters, each with an individual gain slider that allows you to boost or attenuate the given frequency. Filters are spaced at octaves or at one-third octave intervals. Each

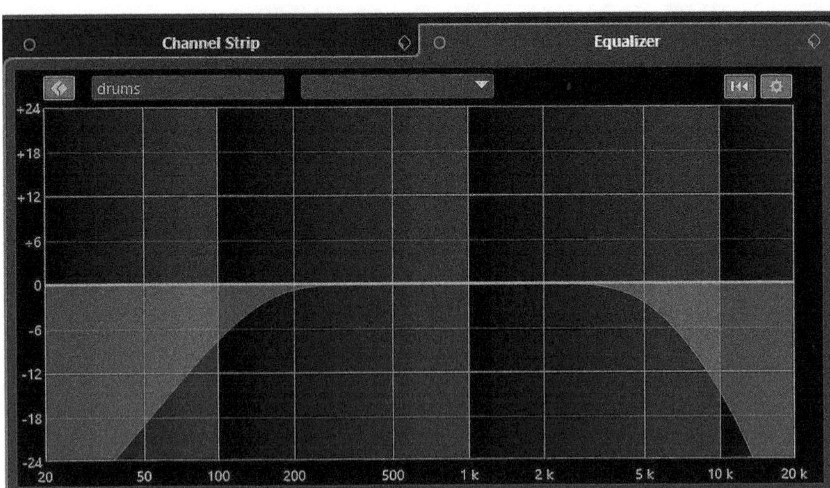

Figure 6.5 Steinberg Cubase high-pass and low-pass filter plug-ins (Courtesy of Steinberg Media Technologies GmbH, a division of Yamaha Corporation, www.steinberg.net)

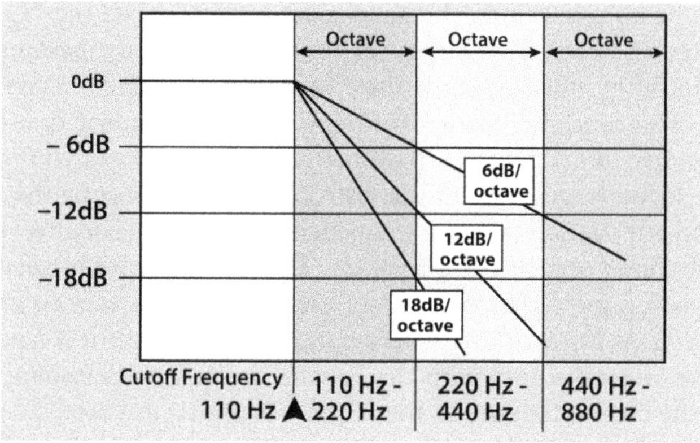

Figure 6.6 Filter slopes

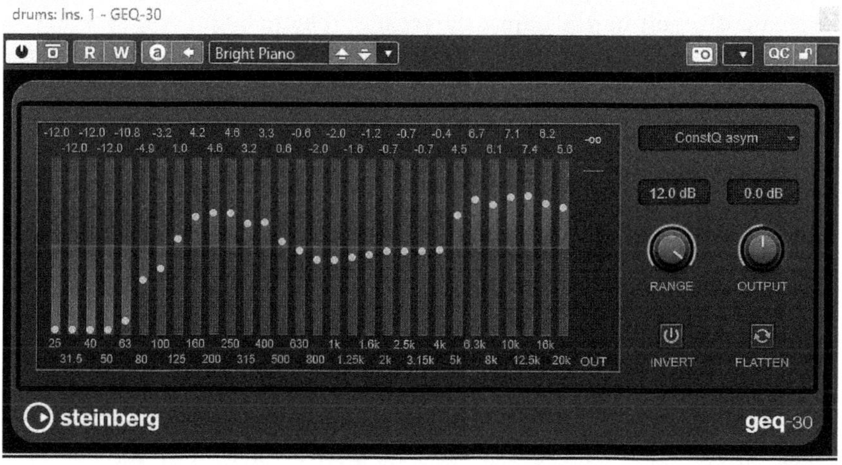

Figure 6.7 Steinberg Cubase graphic equalizer plug-in (Courtesy of Steinberg Media Technologies GmbH, a division of Yamaha Corporation, www.steinberg.net)

filter has a fixed slope as well. Graphic equalizers allow for fast operation because all the frequencies are set and controlled by a single slider. Graphic equalizers do not allow you to specify an exact frequency meaning that sometimes you must work with the surrounding frequencies in order to affect the signal. If you need precise control over frequencies, then you need to use a parametric equalizer (Figure 6.7).

Parametric equalizers offer the most amount of flexibility. Instead of having a boost or cut on either side of the cutoff frequency, a parametric equalizer has a bell-shaped curve on either side of the selected frequency. The width of the bell can be adjusted using the bandwidth control called the "Q." You can boost or attenuate the selected frequency. The flexibility of a parametric equalizer is enhanced by offering you either three or more bands of frequencies to work with. This is useful when working with an instrument that has a wide frequency range, like a piano. A parametric equalizer allows you to attenuate some of the lower frequencies of the instrument while boosting the middle frequencies and attenuating the high frequencies. Parametric equalizers allow you to shape the sound precisely as you want. This flexibility means that you must take care when using a parametric equalizer because it is easy to get lost with all the controls. When working with a parametric equalizer it is a good idea to compare the original sound with the equalized sound to determine if you are improving the overall balance of the sound or making it worse.

Some parametric equalizers feature a real-time display of the frequencies passing through the effect. As you adjust frequencies with the equalizer, you can see how the sound is altered. This immediate feedback is useful when learning to use a parametric equalizer. However, it is always best to listen to the sound carefully as you adjust the settings (Figure 6.8).

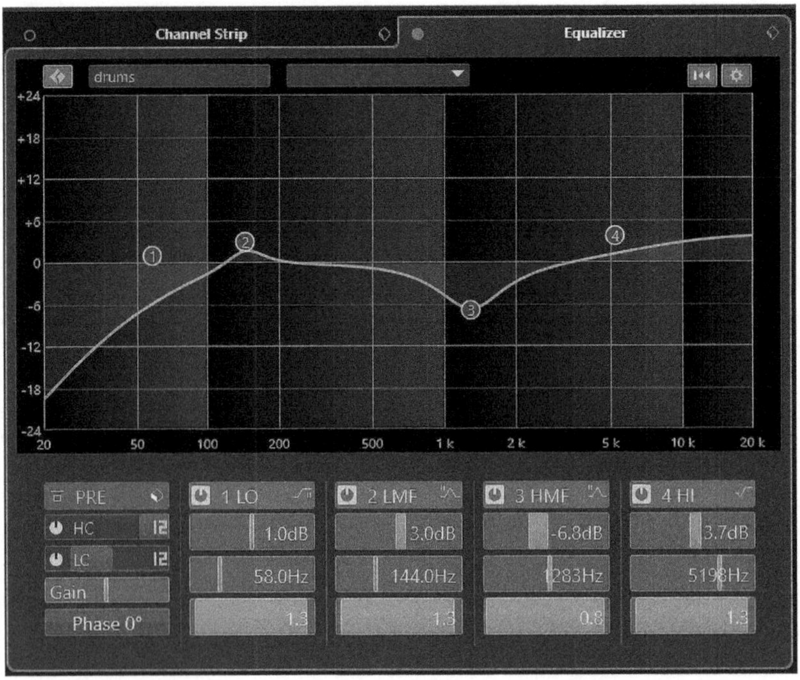

Figure 6.8 Steinberg Cubase parametric equalizer plug-in (Courtesy of Steinberg Media Technologies GmbH, a division of Yamaha Corporation, www.steinberg.net)

Time Processors

Time processors alter the amount of time it takes for sound to reach our ears. Time processors can make audio sound distant or as if they were reflecting off the surface. They can also create sweeping effects that make audio sources sound as if they are moving. These effects were originally achieved using analog tape machines. An engineer would record an instrument like an electric guitar on two tape machines simultaneously. Instead of playing back the two machines at the same time, the second machine would start later by 50 milliseconds or less. The result was an effect of the sound being doubled. If the start time was delayed more than 100 milliseconds, then an echo effect was created. Engineers discovered that they could feed the recorded signal on one machine back into itself to create multiple echoes, or delays. The speed of the tape moving across the play and record heads determined the delay time.

The delay effect creates single or multiple echoes of the original sound. A delay effect in a DAW has four parameters: delay time, feedback, wet amount, and dry amount. The delay time determines how long before the first echo is heard and is measured in milliseconds. DAWs allow you to synchronize the delay time to the tempo of the sequence, making it easy to create delays that are in time with the music. The feedback allows you to specify the percentage of the delayed signal that is fed back into the delay processor. The feedback control allows you to create multiple echoes of the sound. The higher the percentage, the more echoes you will hear. The dry control determines how much of the original signal is heard while the wet control determines how much of the delay you hear. You can balance these two parameters until you achieve a desired sound (Figure 6.9).

If you set the delay time very low, between 15 and 50 milliseconds, you can create a doubling effect where it sounds like there are two instruments playing. This gives the voice or instrument a fuller sound. This is still a popular technique on vocals and guitars.

Engineers also discovered that altering the speed of the tape machine slower or faster while a sound passed through raised and lowered the pitch slightly. These slight changes in pitch made the sound thicker and created a sense of motion. This effect is known as chorus, a modulation effect created using delays as a foundation.

Modulation effects are created when the delayed signal is changed in pitch by altering the delay times by small amounts. The simplest of the modulation effects is vibrato. There are two controls for vibrato: rate and depth/width. Rate determines the speed at which the delay time is altered. The delay time is altered by a low-frequency oscillator (LFO) using a sine wave. As the sine wave goes up, the delay time is increase and then decreased when the wave goes down. This control is measured in hertz and ranges from 0.1 Hz to 5.0 Hz. The depth or width control sets the intensity of the pitch modulation. The higher the percentage, the more pronounced the effect. Vibrato

Figure 6.9 Steinberg Cubase stereo delay plug-in (Courtesy of Steinberg Media Technologies GmbH, a division of Yamaha Corporation, www.steinberg.net)

creates pitch modulations ranging from the subtle to the extreme and can add variation to an otherwise lifeless sound (Figure 6.10).

A phaser begins by taking the original signal and sending it through an all-pass filter which changes the phase of the audio. Combining the original signal with the phased signal will cancel certain frequencies. The LFO controls how much the audio is phased as it passes through the all-pass filter. The feedback controls how much of the phased sound is sent back into the delay circuit. A mix control is added to the phaser allowing you to determine how much of the original and effected sound you hear. The result is swooshing or "jet" sound. Phaser effects are popular with electric guitarists and electric pianists (Figure 6.11).

Chorus is another modulation effect. The chorus starts off as a vibrato effect but adds a delay control. The delay parameter determines the amount of delay in milliseconds added to the original signal. The delay amount is then altered by the rate of the oscillator. A mix control allows you to blend the original and chorused sound. Chorus produces a subtle sweep in the sound that makes it sound doubled. It has more movement than the doubling effect we discussed earlier because of the pitch changes. Chorus is also popular with electric guitarists and electric pianists (Figure 6.12).

Figure 6.10 Steinberg Cubase vibrato plug-in (Courtesy of Steinberg Media Technologies GmbH, a division of Yamaha Corporation, www.steinberg.net)

The flanger effect creates a metallic sweeping sound that is often used on electric guitars. The flanger shares the same controls as a chorus. A feedback control is added so that the effected sound can be fed back into the delay circuit and then mixed with the original sound. Feeding the delayed sound back into the circuit creates phase cancellations that add peaks and dips to the signal's frequency that are constantly changing. These cancellations create the metallic effect that is associated with a flanger. The amount of delayed signal added to the original signal is determined by the feedback control. The higher the feedback value, the more metallic the final sound (Figure 6.13).

Early modulation effects were mono, meaning that the output sound was a single channel. Later models added two outputs for stereo sound. DAW manufacturers such as Steinberg, add a spatial parameter to their modulation effects to increase the stereo presence of the effect. The control is measured in percentage with 100% being the widest stereo effect.

Figure 6.11 Steinberg Cubase phaser plug-in (Courtesy of Steinberg Media Technologies GmbH, a division of Yamaha Corporation, www.steinberg.net)

Figure 6.12 Steinberg Cubase chorus plug-in (Courtesy of Steinberg Media Technologies GmbH, a division of Yamaha Corporation, www.steinberg.net)

Figure 6.13 Steinberg Cubase flanger plug-in (Courtesy of Steinberg Media Technologies GmbH, a division of Yamaha Corporation, www.steinberg.net)

Reverberation

Reverberation, or reverb for short, is the persistence of sound after the sound has stopped. Some people will confuse reverb and delay. Delay is a distinct echo or echoes after the sound has stopped. Reverb is the presence of the sound as ambience. The ambient sound that we hear is what allows us to determine if we are in a tiled bathroom, an underground parking garage, or a concert hall. Even though reverberation is a time processor, I have separated it from the others because of the complexity of the effect.

Prior to digital technology, reverb was added to sound naturally. Acoustic instruments, such as the piano, were recorded in concert halls. You would place a couple of microphones near the inside of the piano to capture the instrument. You would also place a couple of microphones in the hall to capture the ambience of the piano in the room. In the studio, you can blend these two signals to create the amount of ambience you want. Some recording studios would have a special chamber with a pair of speakers and microphones in the space. The sound that they wanted to add reverb to would be played through the speakers and the microphones would capture the ambience of the chamber.

Reverberation is difficult to recreate artificially because the sound is made of up multiple, random, and blended reflections of a sound. Reverb is made up of echoes, but there are thousands of echoes each coming in at different times from different directions. Creating reverb digitally, either as a hardware device or a plug-in, requires an enormous amount of processing power that is not always practical. To overcome this challenge, developers need to be creative and inventive.

The reverb plug-in included with your DAW consists of a series of short delays that randomly select delay times. Each delayed signal is passed through either a low-pass or high-pass filter so that the delayed sound is not identical to the original. Some manufacturers add a chorus to the signal path so that the pitch of the delayed signal is slightly altered, adding to the randomness of the delayed sound. The more variation a plug-in can introduce, the more natural the result of the digital reverb.

To increase the realism of the sound, manufacturers added parameters to shape the sound and create the illusion of a physical space. The first control is pre-delay. Pre-delay determines the amount of time in milliseconds before you begin to hear the reverberated signal. Longer times give the impression that the room is larger and while shorter times will make the room smaller. Pre-delay represents how long or deep the room is.

The early reflection control determines how many reflections are heard after the initial sound. If you are in a small, tiled room like a restroom, you will hear many reflections almost immediately. You will also hear reflections in a small, carpeted room, but they will be fewer and quieter. The early-reflection parameters not only help us represent how wide or high the room is but also help determine the type of material that is used on the walls and ceiling.

The decay time parameter controls how it takes before all reverberation ceases. This is the time it takes before we no longer hear the ambience. Decay time works with the pre-delay and early reflections to create the overall impression of the space you are in. Decay time alone cannot determine the size of the room. A small, tiled, and reflective room could have decay time like a large concert hall. Combining the other two parameters helps create the illusion.

Most reverb effects will include a mix or a dry/wet control that allows you to blend the original signal with the effected signal. More complex reverb effects will give you more control in an attempt to re-create an acoustic space. Some plug-ins, such as the one below by Steinberg, add a room size control that helps define the overall space of the room (Figure 6.14).

Equalization controls are often included with reverb plug-ins so that you can control which frequencies reverberate more than others. Adding reverb to a piano may increase the number of low frequencies in your mix. Removing those frequencies from the reverb helps keep the sound natural. Reverb

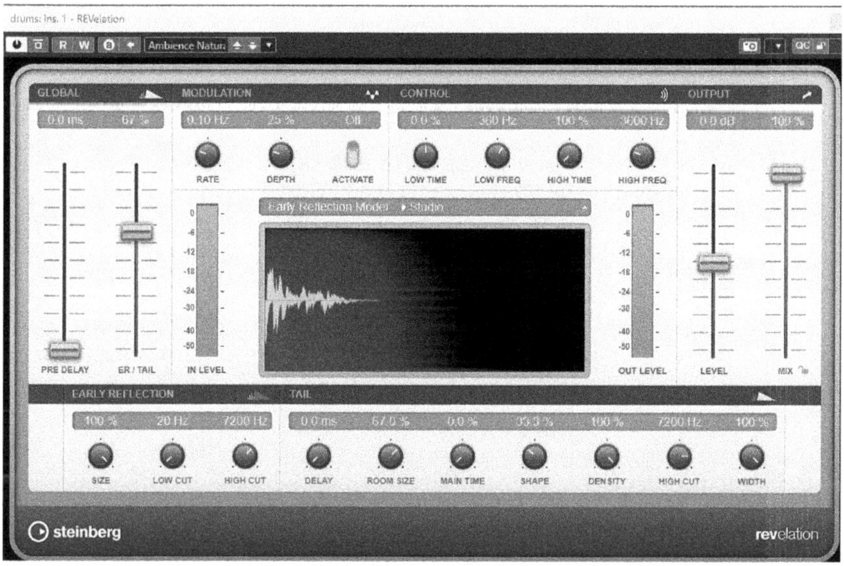

Figure 6.14 Steinberg Cubase REVelation reverb plug-in (Courtesy of Steinberg Media Technologies GmbH, a division of Yamaha Corporation, www.steinberg.net)

plug-ins vary in capabilities and price. It is very easy to fall into the spell of purchasing every reverb you find in the hopes of finding the perfect one for your music. Spend some time with the reverb plug-ins included with your DAW; they may not be perfect, but they may provide you with enough options to give your music the desired ambience.

Summary

In this chapter, we covered effect plug-ins that may be included with your DAW software. We covered the effects families and explored each type in some detail. As you develop your skills, you may discover that there are other nuances to the effects that I did not mention in this chapter. I have provided you with basic information so that you understand the principles of each effect. How you deploy these effects depends on you. Experiment with them and see what you come up with. You will soon discover that some reverb plug-ins offer more parameters than the ones I mentioned. This is expected since every manufacturer has a different approach to developing this effect. When in doubt, consult the user guide for your plug-in to learn the controls. Do not be afraid to experiment. You may surprise yourself with the results.

Activity 6

Effects Processing

In this activity, we will use Audacity to add effects to different sounds. This will give us a better understanding of how effects can alter sound.

Please go to the companion website and download the three WAV files for Activity 6. Please understand that this is copyrighted material and cannot be used for commercial purposes.

Before launching Audacity, create a folder on your computer, either on the desktop or in your documents folder. Name the folder *audio effects* and place the downloaded audio files in that folder. Now we can begin the activity.

1 Launch Audacity
   a From the menu, go to File – Open
   b Go to the location of folder you created
   c Select the *drum-loop* file and click Open
   d Use the magnifying glass with the plus sign in it to zoom in as needed
2 Go to File – Save Project – Save Project As...
   a You will get a warning that a project does not create audio files, click OK
   b Select the project in the folder you created earlier
   c Name your project *audio effects*
   d Click Save
3 Press the Play button to listen to the drum loop. Notice that is a full sound. Now let us listen to the effect of a high-pass filter.
   a Move your mouse over the audio file until it turns into a hand near the top
   b Click the audio file with the hand
   c Go to Effect – EQ and Filters – High-Pass Filter
      i Set the frequency at 500 Hz and the Roll-off slope at 12 dB
      ii Press Preview to hear the results
      iii Notice how the kick drum is no longer heard
      iv Click the X on the top right to close the window
4 Now let us listen to the effect of a low-pass filter
   a Go to Effect – EQ and Filters – Low-Pass Filter
      i Set the frequency at 1,000 Hz and the Roll-off slope at 12 dB
      ii Press Preview to hear the results
      iii We can hear the kick drum, the high hat sounds muffled
      iv Click on the X on the top right to close the window

5 Now we will explore a compressor

   a Go to Effect – Volume and Compression – Compressor
      i Set the Threshold at -10 dB
      ii Leave the Noise Floor at -40 dB
      iii Set the ratio to 3:1
      iv Leave all the other parameters at their default setting
      v Uncheck the box that says Make-up gain for 0 dB after compressing
      vi Press Preview and listen to the results
      vii The drums are quieter, but still at a good volume

   b Now click on Apply
      i The waveform is smaller, but notice that parts of it are still louder than others
      ii Go to Edit – Undo Compressor

6 Now let us listen to a Phaser

   a Go to Effect – Distortion and Modulation– Phaser
      i Set the Stages to 6
      ii Set the Feedback to 30%
      iii Leave all the other parameters at their default setting
      iv Press Preview and listen to the results
      v Notice how the drums sound

   b Experiment with different Feedback settings to get more dramatic results
   c Click the X at the top right to close the window when you are done

7 Let us hear what a Delay does to our drums

   a Go to Effect – Delay and Reverb – Delay
   b This time, play with the various settings on your own to hear what a delay does to the drums
   c When you are finished, click on the X at the top right to close the window

8 The final effect we will look at is Reverb

   a Go to Effect – Delay and Reverb – Reverb
   b Click on the Preview button to listen to the default reverb settings
   c Experiment with the settings and see if you can put the drums in a small room and a big room

9 You can continue with the Activity by using the guitar and bass files to trying out different effects on them.

This concludes this activity. If you want, you can save the project or quit without saving.

Terms

Chorus An effect that produces a subtle sweep in the sound that makes it sound doubled.

Compressor An effect that reduces the dynamic range of an incoming signal

Cutoff Frequency The frequency at which an equalizer begins to affect the incoming sound.

Delay An effect that produces single or multiple repetitions of a sound.

Dynamics Processor An audio effect that alters the dynamic range of a sound

Filter Slope The rate or angle that a filter boosts or attenuates a signal after the cutoff frequency.

Flanger An effect that produces metallic sounds.

Gate An audio effect that only allows sound to pass if it is above a certain volume level.

Graphic Equalizer An effect that uses regularly spaced, fixed frequency filters, each with an individual gain slider that allows you to boost or attenuate the given frequency.

High-Pass Filter An audio effect that allows frequencies above a specified frequency to pass and be heard.

Limiter An audio effect that limits the loudness of an audio signal being output.

Low-Pass Filter An audio effect that allows frequencies below a specified frequency to pass and be heard.

Parametric Equalizer An audio effect that allows you to boost or attenuate specific frequencies of a sound.

Phaser An effect that creates a sweeping sound by altering the harmonics of a sound.

Reverb An effect that creates the persistence of sound after the initial sound has stopped.

Shelving Filter An effect that boosts or attenuates all signals above or below a selected frequency.

Vibrato An audio effect that slightly alters the pitch of a sound at regular intervals to create fluctuations in pitch.

Chapter 7

Audio Hardware

Audio Hardware

In this chapter, we will look at the type of audio hardware you might need to start making music in your home studio. We will go over the basic functions of audio hardware and their functions. This chapter is meant to give you a broad overview of audio hardware options and their purpose. For those of you wishing to gain a deep understanding of audio hardware, please refer to the companion website for suggested reading and additional resources.

The type of hardware you need depends on the type of music you intend to produce and your workflow. To aid our conversation, I will set a baseline configuration of two inputs for microphone and instruments, two outputs for speakers, a keyboard controller, a pair of speakers, and a pair of headphones. This baseline configuration provides flexibility and room for growth. Since products evolve and manufacturers add new products, we will not look at specific brands. Rather, we will learn to identify the features we need to get started, and how to determine if a product meets those needs.

Many companies advertise that their product will help achieve better mixes and professional results. Do not fall into the trap that newer and more expensive is better. As I stated earlier in this book, I know musicians using old hardware, old computers, and old software, and creating and achieving professional results. It is not the hardware that determines the quality of sound that you create but rather how you use that hardware. Understanding the limitations of your budget and the capabilities of the hardware you are purchasing will make an enormous difference to your music productions.

Audio Interfaces

Along with your computer, the audio interface is an essential component of your studio. The audio interface has an analog-to-digital converter (ADC) to digitize sound received at the input. The sound is then sent digitally to the computer for processing and storage. The digital-to-analog converter (DAC) converts the digital data stored on your computer to electrical energy that

DOI: 10.4324/9781003345138-8

is sent from the outputs of the interface to your speakers or headphones. The quality of the ADCs and DACs is one variable in the cost of an audio interface. More expensive models will offer higher-quality converters. However, entry-level interfaces offer quality converters suitable for all recording situations.

Audio interfaces vary in the number and type of inputs and outputs they accommodate as well as how they connect to your computer. Let us examine the various components of an audio interface so that we can decide what features are important and necessary for our work.

Computer Connection

Many entry-level interfaces connect to your computer with a USB type A or type C cable. Several manufacturers are generous and the USB cable with your purchase. Many USB audio interfaces are bus powered, meaning that they receive power over the USB cable and do not require a power adapter. USB connections are found on all Windows and Apple computers.

Thunderbolt is another option, but this is usually found on higher end models that provide many inputs and outputs. Thunderbolt interfaces require that your computer has a Thunderbolt port, which is common in Apple computers, but harder to find in Windows PCs.

Inputs

Entry-level audio interfaces feature at least two combo jacks. A combo jack is a connector that can accept either XLR (microphone) or 1/4-inch (instruments) cables. The audio interface automatically detects whether an XLR or 1/4-inch cable is connected. When connecting a guitar or bass directly to the jack, you usually need to activate a button on the device to accept those inputs. Guitar and bass instruments are high-impedance instruments whereas instruments such as synthesizers are low impedance. The audio interface inputs are low impedance by default; the switch allows the input to accept high-impedance instruments. A practical definition of impedance is the resistance to electrical current. Impedance is represented by the letter Z. Your interface might have a button labeled Hi-Z, which when activated, allows you to connect high-impedance instruments (Figure 7.1).

If you need more inputs, some models will add two, line-level 1/4-inch jacks, giving you a total of four inputs. However, the additional jacks are strictly line level, so you cannot connect a guitar or bass to those jacks. Models that feature four combo jacks offer you greater flexibility. If you plan on recording a drum set or multiple instruments and vocalists at once, then a model with eight combo jacks might be required. The number of jacks required depends on your workflow and needs. Many of us who work with virtual instruments and occasionally record vocals or guitars can easily work

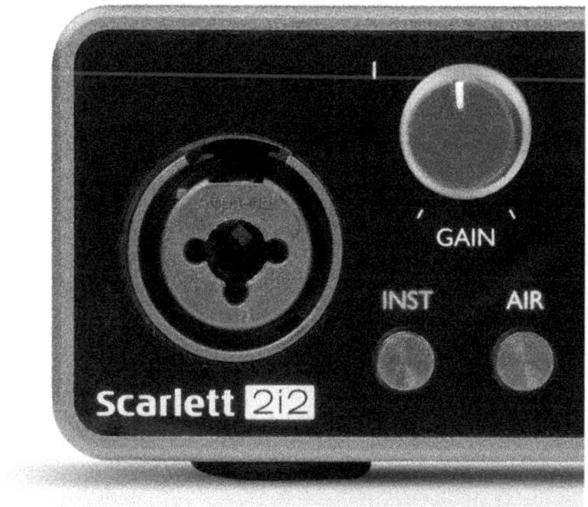

Figure 7.1 Image of combo jack input on the Focusrite Scarlett 2i2 (Courtesy of Focusrite Plc, www.focusrite.com)

with two combo jacks. Keep in mind that adding more inputs will increase the price of the interface.

Preamplifiers

All audio interfaces include a preamplifier for their combo inputs. The preamplifier raises the voltage or gain of the incoming microphone signal up to line level, which is the level the ADC in the audio interface expects. The preamplifier can raise the gain of instrument, line, and microphone signals. The amount of gain that a preamplifier can provide is measured in decibels. A preamplifier with a gain of 60 dB can increase the strength of an incoming signal by 60 dB.

The audio interface will have a signal light or meter that tells you the level of the incoming signal. As you increase the gain knob on the amplifier, the signal level will increase. As the signal gets stronger, the meter or light will turn green. When it is too loud for the circuitry in the audio interface, the meter, or light turns red. The yellow and orange lights are in between the green and red. Manufacturers recommend setting the level slightly above green or when the light turns yellow.

The quality of the preamplifiers affects the cost of the audio interface. One variable is the amount of electronic noise present in the circuitry. Another variable is the transparency of the preamplifier; some slightly change the color or tone of the sound. The ability to hear the subtle differences in preamplifiers depends on your hearing and experience. At this point in your journey, focus on making music, and less on whether the quality of your preamplifiers is the same as those in a professional recording studio.

Outputs

An audio interface will have at least two 1/4-inch outputs. The volume knob at the front of the interface controls the output level. The jacks accept balanced and unbalanced cables. For most users, two outputs are sufficient to connect one pair of speakers. Some models offer four 1/4-inch outputs, allowing you to connect another set of speakers or route audio to another location. Models aimed at DJs will have a pair of Radio Corporation of America (RCA) jacks for easy connections to consumer systems. If you need even more outputs, there are models that have eight 1/4-inch outputs.

The number of outputs depends on your workflow. If you plan on working in surround sound, then additional outputs are required. You also need to consider that the level of the additional outputs may not be controlled by the main volume knob. This means if you connect a set of speakers to the additional outputs, you will need to control their level directly from the speaker. If you work in stereo and have one set of speakers, then a pair of outputs is all you need.

All interfaces will also offer at least one headphone output. The headphone jack is a 1/4-inch jack. If you plan on connecting earbuds to that jack, you will need an 1/8-inch to 1/4-inch adapter. The headphone jack has its own volume control, separate from the main volume. Some models will offer two headphone outputs, which is handy if you are working with another person, and both want to use headphones.

MIDI Connections

MIDI stands for Musical Instrument Digital Interface. MIDI is a protocol that allows MIDI devices such as synthesizers and drum machines to communicate with each other and computers. A computer needs a MIDI interface to communicate with MIDI devices. Many audio interfaces offer a MIDI IN port and MIDI OUT port. If you plan on working with external synthesizers and drum machines in the future, then you may consider purchasing an audio interface with MIDI ports. Many entry-level models include MIDI ports. We will look at MIDI in detail in Chapter 8.

S/PDIF

There are audio interfaces that have S/PDIF or optical digital connections. S/PDIF stands for Sony/Philips Digital Interface. This digital standard was developed by Sony and Philips for CD players to connect to consumer devices digitally. S/PDIF uses either coaxial cables with RCA connectors or optical cables with TOSLINK (Toshiba Link) connectors. The connection is one-way only, so you need two cables to send information back and forth between the two devices.

The S/PDIF protocol can carry two channels of uncompressed audio or six channels of compressed audio for 5.1 surround sound. The S/PDIF connections on audio interfaces are generally configured for stereo audio. Unless you have equipment that uses S/PDIF, there is no need to purchase an audio interface with these connections.

ADAT

The Alesis Digital Audio Tape (ADAT) was an eight-track digital audio recorder developed by Alesis in 1992. The unit could record eight tracks at a sample rate of 48 kHz and a bit depth of 24 bits on S-VHS tapes. The system was revolutionary at the time and offered home studios an affordable option for digital recording. To allow ADAT machines to exchange audio, Alesis developed the Lightpipe protocol which carries eight channels of digital audio in one direction over a TOSLINK optical cable. As hard disk recording became viable, the popularity of the ADAT machine faded. However, the Lightpipe protocol still exists in many devices today.

An audio interface with ADAT Lightpipe connections can receive eight channels of digital audio from another device. If your audio interface has two microphone inputs and you want to add more inputs, then you need to purchase another audio interface with more inputs. If your interface has an ADAT connection, then you could purchase an eight-channel microphone preamplifier with ADAT connections and expand your inputs.

Digital connections will add to the cost of the audio interface so consider your needs before committing to purchasing an interface with either option.

Choosing an Interface

The type or style of music you wish to create will help you determine the type of interface you should purchase. If you create mainly electronic music with the sounds and instruments in your computer, then an interface with two inputs and two outputs is probably all you need. The inputs are useful if you ever want to add vocals or an external instrument to your music. Songwriters who work with acoustic instruments may want four microphone inputs so they can record a few instruments and a voice at once. If you have an external synthesizer and want to record acoustic instruments or voice, then a model with two microphone inputs and two additional line inputs would be helpful.

The number of outputs also varies with your workflow. If you are a DJ, you may want to send two different mixes across two pairs of outputs connected to your mixer. Or maybe you want the option to use two pairs of speakers. If you are working mainly in stereo with one set of speakers, then two outputs are sufficient.

Spend time researching products and reading user reviews. These reviews are helpful to learn how others feel about a product. Visit the manufacturer's website and download the user manual for the interface. This is the best way to learn the features of the interface. Remember that your first interface should help you get started in music creation. In time, as your needs and experience grow, you may need to purchase a new interface with more options. The quality and options found in entry-level interfaces are very good, so do not feel that you are compromising by not spending a lot of money.

Audio Cables

Audio cables are essential because they allow us to connect instruments, microphones, and speakers to our audio interface. Let us explore the differences between cables.

Quarter-Inch Cables

Quarter-inch cables are used to connect anything that is not a microphone. It is possible to use a 1/4-inch cable on a microphone, but the professional standard for microphones is the XLR cable, which we will look at later. The 1/4-inch cable derives its name from the diameter of the connector. This cable is used to connect line devices, such as synthesizers and speakers, and instrument-level devices, such as guitars and basses. However, 1/4-inch cables can be unbalanced or balanced. To complicate matters even further, there are two standards for how loud line-level signals are.

Line Levels

Signal level refers to how strong or "hot" the electrical energy is when it travels down the cable. Line levels are the strongest levels, meaning that the current coming out of synthesizer or audio interface is high. Higher currents require less amplification when they reach their destination. In the case of a synthesizer connecting to an audio interface, the preamplifier in the interface only needs to increase the current slightly.

Line-level signals operate at two different levels, consumer or professional. The consumer standard used in home studios and consumer equipment is -10 dBV. The professional standard is +4 dBu. The difference between these two levels is not easy to explain since each standard measures voltage differently. What is important to understand is that the voltage level of consumer devices is lower than that of professional ones. The professional standard is roughly 12 dB louder than the consumer standard.

Why does this matter? If you connect the outputs from an audio interface, which is running at +4 dBu into a consumer stereo system, the signal might be too loud for the system to handle. If you connect the output of a consumer CD player into the audio interface, the signal might be too weak. Do you need to be concerned about this possibility? No, it is not a major concern. Audio interfaces are more than capable of handling professional- and consumer-level signals. You just need to be aware that the CD player might require an increase in the input level of the interface to achieve a strong signal.

Unbalanced Cables

When you take apart an unbalanced 1/4-inch cable, you will see two wires underneath the sheath. One wire carries the audio signal and the other the ground. The electrical system in your home must have a connection to the earth to minimize interference from external sources. The ground wire in your audio cable is an attempt to minimize interference from radio, electric, and magnetic sources. If the electrical outlets you are connecting your equipment to are grounded, and the home has a stable ground, then the chance of interference is minimal. However, their ground wire in an unbalanced 1/4-inch cable is still susceptible to interference, even in the best situations. Any noise picked up from the ground is mixed with the audio signal. This noise can add hum or static to your signal (Figure 7.2).

Unbalanced 1/4-inch cables are called Tip Sleeve (TS) cables. The connector has a tip, which carries the signal, and a sleeve, which carries the ground. You can easily identify an unbalanced cable by looking to the connector at the end of the cable (Figure 7.3).

The RCA phono connector is also used with unbalanced cables. The RCA phono connect was developed by the RCA in the 1930s to connect their turntables to amplifiers. RCA phono connections are still used for turntables and consumer stereo equipment.

Balanced Cables

Balanced cables have three wires. The one wire carries the ground. The other two carry a positive and negative version of the audio signal. When the signal reaches the end of the cable, the polarity of the second wire is reversed from negative to positive and the two signals are combined. Any interference noise

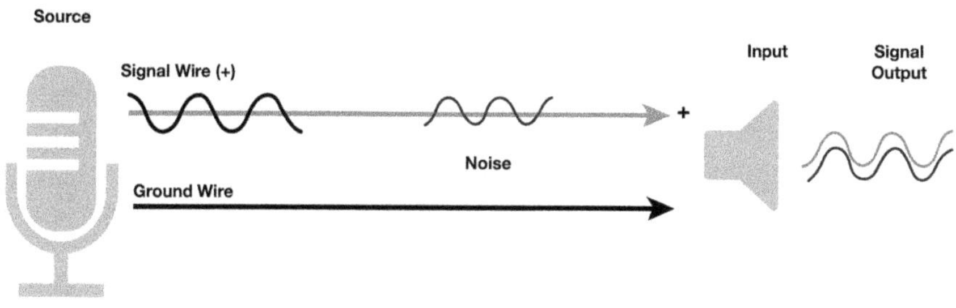

Figure 7.2 Signal path when using an unbalanced cable

Figure 7.3 Image of unbalanced 1/4-inch cable showing tip and sleeve (Courtesy of Hosa Technology, www.hosatech.net)

picked up by the ground wire is transferred to both signal paths at the same polarity. This means that both cables carry positive versions of the noise. When the signal reaches the end of the cable, the noise on the negative signal is inverted to negative. Now, we have a positive and negative version of the noise. We know that signals that are 180 degrees out of phase will cancel each other out, which is what happens to the noise. This means that under ideal situations any hum or interference to the other two wires will be canceled out at the end and you will not hear the hum or interference (Figure 7.4).

Balanced 1/4-inch cables are called Tip Ring Sleeve (TRS) cables. The connector has three points of contact. The tip carries the original signal. The ring carries the inverted signal. The sleeve carries the ground. The TRS cable looks similar to the TS cable except that there is an extra separation for the ring (Figure 7.5).

Given this information, you might be inclined to use balanced cables for your home studio to reduce your chances of interference. Balanced cables are more expensive than unbalanced cables. If your electrical outlets are grounded correctly, then your chances of interference are minimal. No one ever ruined a recording by using unbalanced cables.

Instrument Levels

Instrument levels pertain to instruments such as guitars and basses. The signal level is equivalent to line-level signals and can travel over balanced

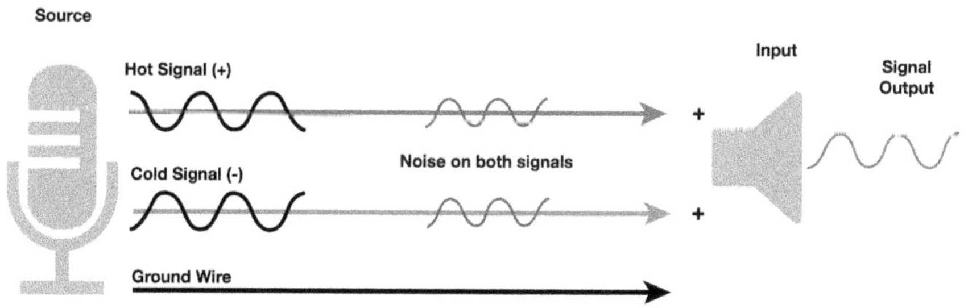

Figure 7.4 Signal path when using a balanced cable

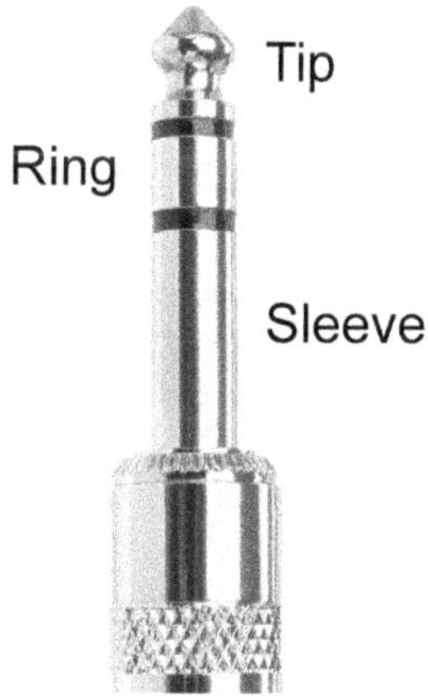

Figure 7.5 Image of balanced 1/4-inch cable showing tip, ring, and sleeve (Courtesy of Hosa Technology, www.hosatech.net)

and unbalanced cables. Instrument level is a high-impedance signal, which we discussed earlier. Your audio interface can accommodate low- and high-impedance signals. There is usually a button next to the input jack that you must press when connecting a high-impedance signal. In the rare occasion that your audio interface does not accept high-impedance signals, you will need a direct injection (DI) box.

Direct Injection

A DI box converts high-impedance signals to low impedance. A DI box also converts unbalanced signals to balanced signals. Finally, it converts a line-level signal, which is strong, to a microphone-level signal which is much weaker. A DI box allows you to connect an electric guitar directly to the microphone input of your audio interface. The DI box is a simple, but extremely useful item to have when you must connect anything to a microphone input.

Microphone Level

Microphone levels are low impedance as well as low level. The design of the microphone is such that the resulting signal level is weak. A preamplifier is required to increase the voltage of the microphone signal so that it can be run through a mixing console or audio interface. All audio interfaces include a preamplifier to boost the signal level. Microphones require balanced connections using XLR cables. The XLR cable was developed by Cannon Electric in the 1950s and uses a three-pin connection with a latching mechanism to prevent the cable from accidentally disconnecting. The three pins connect to three wires, which allows for balanced connections (Figure 7.6).

Microphones come in a variety of designs and features, which we will explore later in this chapter.

Latency

Latency is a term that is often mentioned when manufacturers advertise their audio interfaces. Latency is the time it takes a signal entering the inputs of an audio interface to be processed by the computer and sent back the outputs of the interface. When you sing into a microphone or play an instrument connected to the audio interface input, the audio is digitized and then sent to your computer for processing. The computer stores the digital data on the hard drive. The computer then sends the data back to the audio interface for

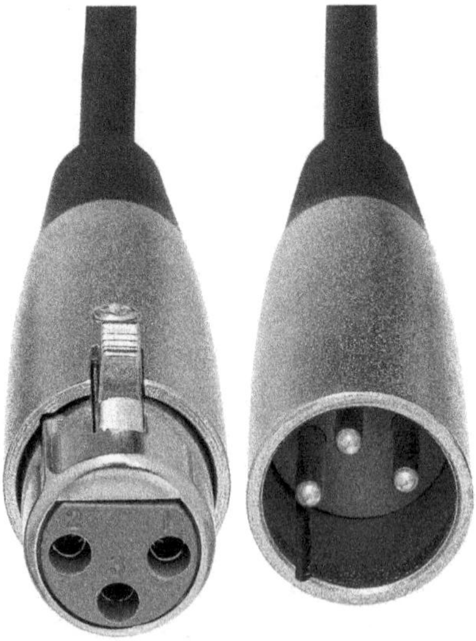

Figure 7.6 Image of XLR cable showing the connector and three pins (Courtesy of Hosa Technology, www.hosatech.net)

conversion to an electrical signal which drives your speakers or headphones. This process is not instantaneous. If you hear an echo of yourself after creating a sound, you are experiencing latency. The image illustrates the process in Figure 7.7.

To ensure accurate audio sampling, DAW software imposes a buffer that controls how much data is collected before being sent to the computer. The buffer prevents the computer from overloading by controlling how much information is sent at once. A large buffer decreases the computer load by collecting more data, but this adds latency. A smaller buffer reduces latency but requires the computer to process the audio immediately. If the computer cannot keep up with the data stream, it will drop data which leads to audible errors such as clicks and crackles in your audio. Smaller buffers require faster computers.

Multicore processors in computers allow them to be faster and more efficient, thus reducing dropouts and errors when running low buffers. However, even under ideal situations, you will experience latency regardless of how fast your computer is or how low you set the buffer. Fortunately, all audio interfaces, even the entry-level ones, offer a direct monitoring solution to reduce the amount of latency that you experience. Direct monitoring splits the incoming audio signal. The first signal goes through the ADC in the audio interface and is sent to the computer. The second signal is routed to the headphones or speakers so that you immediately hear yourself. However, what happens if you are recording yourself and using tracks already on your

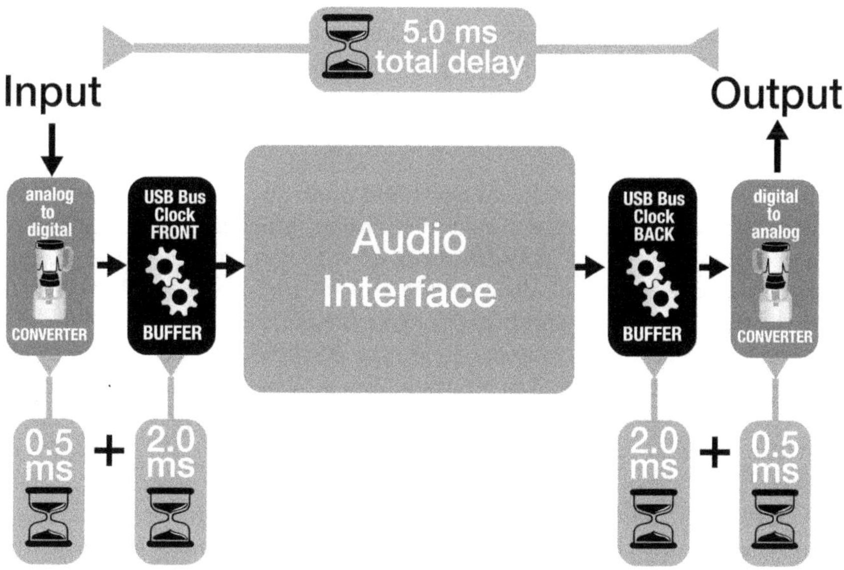

Figure 7.7 Image showing signal flow to a computer showing delay times

computer? Would there be latency since the tracks on the computer still need to be converted to analog?

The buffer size controls the rate at which data enter and exit the computer. When you press play on your DAW, the computer will buffer data before any sound is heard. Doing so ensures that the sound is being delivered to your headphones and speakers immediately. If you have many tracks, each with plugins, you might experience a slight delay between the time you press play and the time you hear the music. This delay is caused by the buffer; it needs to gather enough data before beginning playback. The DAW will wait until there is enough data in the buffer before beginning playback so that there is no delay when you are recording new material against existing tracks.

Direct monitoring is handled differently by each manufacturer. Some enable the feature automatically, so you never have to think about it. Others have a knob on the front that lets you control how much audio you hear from the computer and how much from your input. Some companies give you more control of what you hear using the software control panel for the interface. Remember that direct monitoring may not remove all latency, but it will reduce the delay enough so that it is barely noticeable.

Microphones

Microphones are transducers; they convert one form of energy to another. A microphone converts acoustic energy, which is moving air molecules into electrical energy. All the microphones consist of a membrane that vibrates in response to the movement of air molecules and converts those vibrations into electrical energy. We will look at three different types of microphones, how they capture sound, and their tonal characteristics.

Dynamic Microphones

Dynamic microphones are the simplest and most durable of the three types. The primary components are a mylar diaphragm, which is a flat disc, attached to a voice coil, which is a cylindrical coil of wire and a magnet. The magnet slides inside the voice coil allowing the coil to freely move back and forth over the magnet. When sound waves strike the diaphragm, it moves the coil back and forth across the magnet creating a small electrical current through electromagnetic induction. The electrical current is weak, but strong enough to travel along the XLR cable until it reaches the preamplifier of the audio interface. Dynamic microphones pick up sound from the top of the capsule (Figure 7.8).

The primary components are housed in the capsule, which looks like a cylindrical basket. The overall design of a dynamic microphone is sturdy. You can feel the heft of the components when you pick up a dynamic microphone. These microphones can handle the abuse of live sound and will survive being

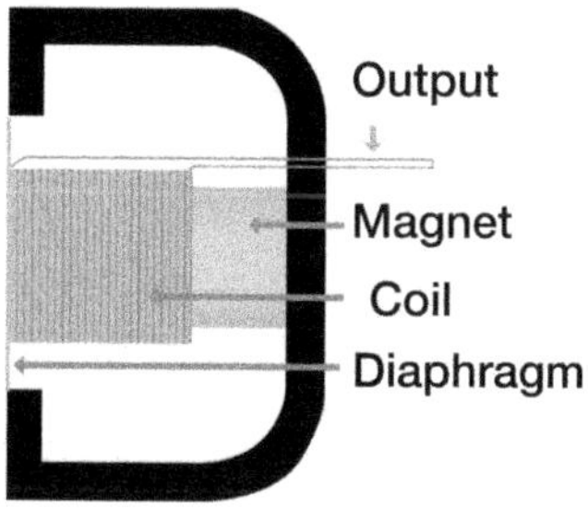

Figure 7.8 Image of interior of dynamic microphone

dropped. Dynamic microphones make excellent first microphones because they are durable and can handle extremely loud sounds. The voice coil can be slow to respond so this microphone does not always respond quickly and is not sensitive enough for quiet instruments. However, the simple construction of a dynamic microphone makes them affordable.

Ribbon Microphones

Ribbon microphones share a similar design to dynamic microphones. Instead of a mylar diaphragm, a ribbon microphone uses a light metal ribbon. The ribbon is connected to wires on both ends and is suspended in between the positive and negative poles of a magnet. When sound strikes the ribbon, it vibrates and generates electric current through electromagnetic induction. The change in current travels along the wires to the XLR cable and to the audio interface. Like the dynamic microphone, the signal is weak and requires a preamplifier. Ribbon microphones pick up sound from either side of the capsule (Figure 7.9).

Older ribbon microphones are very fragile and require careful placement so that the ribbon is not damaged by loud sounds. Modern ribbon microphones are much more durable and can withstand loud sounds such as kick drums and electric guitar amplifiers, however, one should take care that they are at least one foot away from the source. Ribbon microphones are extremely sensitive and can easily pick up quiet sounds and subtle changes in dynamics. The sound from a ribbon microphone is smooth and mellow. Ribbon microphones are expensive compared with dynamic microphones.

Condenser Microphones

Condenser microphones are the most versatile of the three types. Condenser microphones use an electrostatic field to capture sound. The field is created by two thin metal plates placed very close to each other. The top plate, which is usually coated with gold to better conduct electricity, moves back and

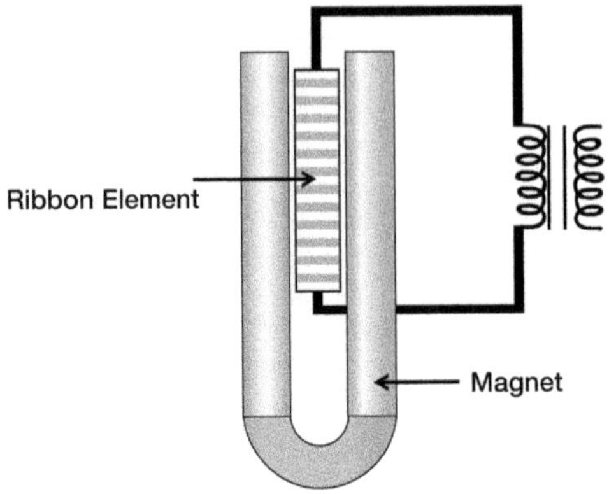

Figure 7.9 Image of interior of ribbon microphone

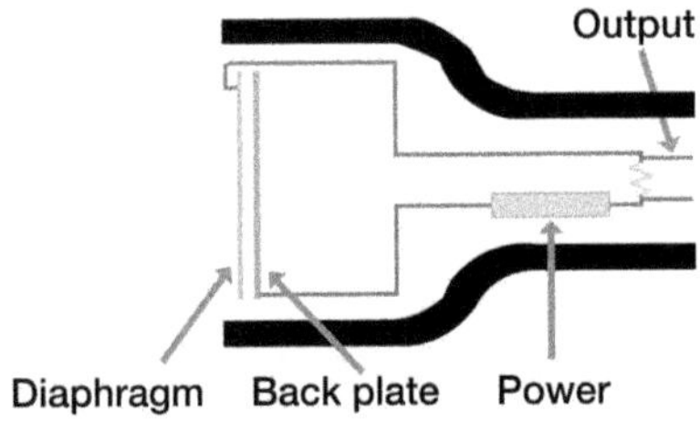

Figure 7.10 Image of interior of condenser microphone

forth, behaving like a diaphragm. The second plate is fixed. Each plate has a thin wire connected to it carrying an electrical charge. The charge creates an electrostatic field between the plates. When sound strikes the top plate it vibrates, the current in the field changes. This change in current is sent across the wires, down the XLR cable to the audio interface. The signal from a condenser microphone is stronger than that of a dynamic or ribbon microphone but still requires a preamplifier (Figure 7.10).

Condenser microphones are suited for a variety of audio sources. The electrostatic design allows for small and large designs to accommodate a variety of needs. These microphones are responsive to loud and quiet sources and can pick up subtle details and dynamics in sounds. They are suited for a large variety of sources and some microphones are designed for specific sources and instruments. Pricing for condenser microphones ranges from affordable to expensive, depending on your needs.

Condenser microphones generally pick up sound from the front side of the capsule. The manufacturer logo is located on the front side you know which side is the front. Some manufacturers configure condenser microphones to multiple patterns.

Microphone Patterns

We already know that our ears allow us to hear sound in a 360-degree circle around our head. We can use the same principle when discussing how microphones "hear" sound. What we are describing is the microphone's pickup pattern. We will use a condenser microphone as an example. A condenser microphone picks up sound from the front of the capsule. When the front of the microphone is aimed at the sound source, then the microphone is listening at zero degrees or on axis. If the microphone has the back side aimed at the source, then it is 180 degrees off axis or rear axis. Sounds aimed at either of the smaller sides of the microphone are 90 degrees or 270 degrees off axis or side axis.

Omnidirectional

A microphone that can capture sound equally across 360 degrees has an omnidirectional pickup pattern. These microphones are ideal for capturing the ambient sound of a room or a group of singers in a circle around the microphone. Dynamic and condenser microphones can be omnidirectional, depending on the manufacturer. It is best to not assume that all dynamic and condenser microphones are omnidirectional. The product specifications will tell you the microphone's pickup pattern (Figure 7.11).

Bidirectional

Bidirectional patterns are all called figure-eight patterns. This pattern picks up sound equally at the front and rear axis. The quality of the signal is identical from either side. However, most of the sound is rejected from the sides.

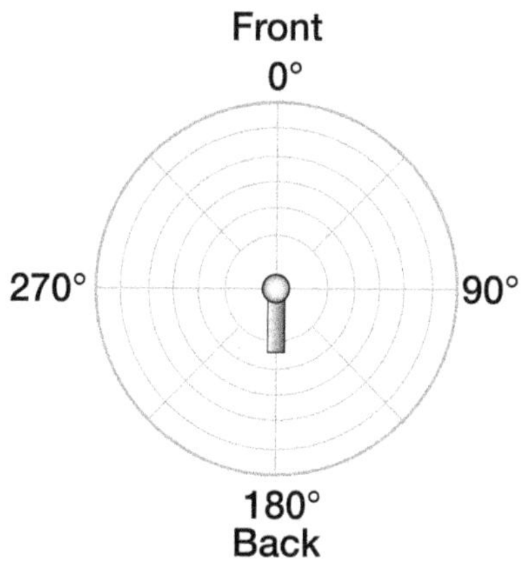

Figure 7.11 Image of omnidirectional microphone pickup pattern

This microphone pattern is ideal if you need to pick up the front and rear of a room but not the sides. It is also well-suited if you have two singers and only one microphone input available. All ribbon microphones are bidirectional as are some condenser microphones (Figure 7.12).

Cardioid

The cardioid pattern is after its shape, which resembles a heart. This pattern mainly focuses on the front of the microphone. Some sound can be captured from the sides. The rear of the microphone picks up very little sound. Microphones with this pattern are well-suited for situations where you need to capture all of what is in front of the microphone and less from the sides and rear. Nearly, all condenser microphones have a cardioid pickup pattern. Many dynamic microphones also have a cardioid pattern (Figure 7.13).

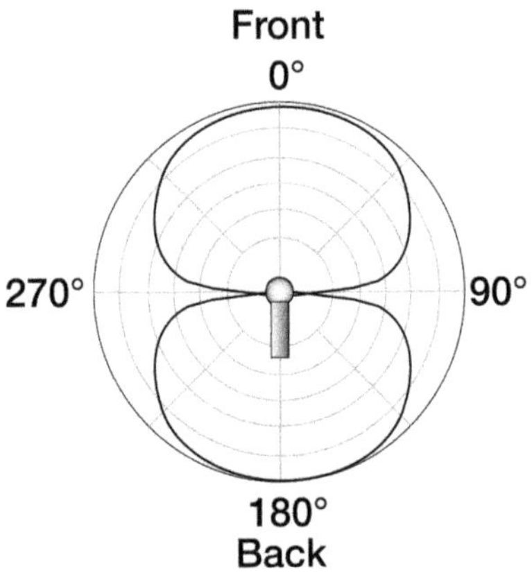

Figure 7.12 Image of bidirectional microphone pickup pattern

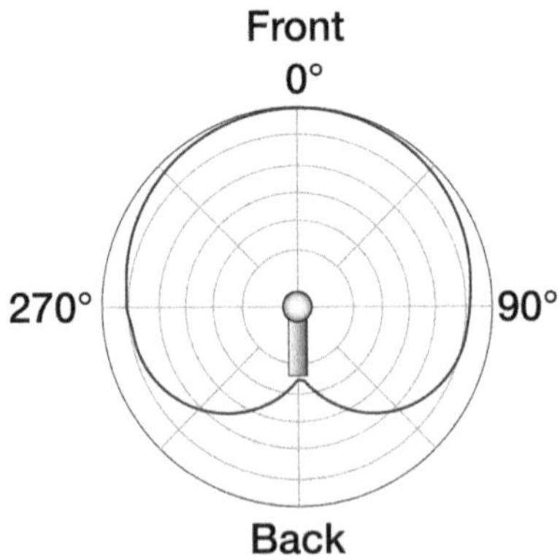

Figure 7.13 Image of cardioid microphone pickup pattern

Multi-Pattern Microphones

The electrostatic design of condenser microphones offers flexibility because they can be designed to have multiple patterns. Some manufacturers offer condenser microphones that have a selector that allows you to choose the pickup pattern for the microphone. These microphones allow you to choose between cardioid, omnidirectional, and bidirectional patterns. The ability to choose a pattern based on the situation is useful. These condenser microphones do cost more than single-pattern microphones, but their flexibility is great value for many users.

Phantom Power

All condenser microphones require an electrical source to generate the electrostatic field. One solution is to place batteries within the body of the microphone to generate power. A more practical and common solution is to send an electrical current to the microphone through the XLR cable. This electric current is called phantom power. Phantom power is 48 volts of current supplied to the microphone on pins two and three of the XLR cable. This power is supplied by the preamplifier in the audio interface. It is called phantom because the power is being fed through the same lines that are carrying the audio to the preamp.

Not all microphones require phantom power, and you should take care when enabling phantom power so that you do not damage your microphones. Dynamic microphones are unaffected by phantom power. If you turn on phantom power when a dynamic microphone is plugged in it ignores the current. Active ribbon microphones require phantom power while passive ones do not. Feeding phantom power to a passive microphone could damage it, so check the owner's manual before enabling phantom power.

If your microphone requires phantom power, follow these steps every time you use that microphone. First, make all the connections between the microphone and audio interface. Second, turn on the phantom power and wait a few moments for the current to reach the microphone. Now, adjust the gain knob on the interface to engage the preamplifier. When you finish your session, turn the gain knob all the way down. Then turn off phantom power. Wait a few moments for the current to discharge before disconnecting the microphone from the cable. Modern microphones are durable and can probably handle errors such as unplugging the microphone before the gain is lowered and phantom power is turned off, but you do want your equipment to last so it is best to always take precautions.

Frequency Response

You will recall that our range of hearing, or bandwidth is 20 Hz–20 kHz. We could also say that the frequency response of our ears is 20 Hz–20 kHz.

Microphones also have a range of frequency response. That frequency response is rarely equal throughout the range; there will be areas of frequency where the microphone is more sensitive than others. Manufacturers include the frequency response of a microphone, but these specifications are generated in laboratories under ideal conditions. Use the specifications as a guide and let your ears make their own judgment.

The frequency response of a microphone will change depending on whether the source is on or off axis. Some microphones are less sensitive in the higher frequencies when they off axis. If you are placing a microphone in front of a guitar speaker cabinet and the tone is too bright for you, placing the microphone slightly off axis might yield a warmer tone.

Microphone Placement

Microphone placement follows the concept of "best practices." This means there are no hard-set rules as to where to place a microphone, but there are certain locations that yield better results. Therefore, use any microphone placement guide you find as a starting point. You should always listen to the sound you are capturing and adjust the microphone until you achieve the sound you want. Avoid the mentality of fixing the issues later in the DAW. If a kick drum does not have enough low end for you, then adjust the microphone until you find the sound you want. It is always best to start with the best sound you can before you start recording.

Spend time experimenting with your microphone and listening to the results. Every microphone has its own characteristic sound and learning how to work with that sound is important. Microphone placement affects how the microphone "hears" the sound, which can work to your advantage when you are trying to achieve certain results. Let us look at other variables to consider before addressing microphone placement best practices.

Proximity Effect

The proximity effect is when the bass response of a microphone increases the closer you place the microphone to the source. This is most noticeable when singing or speaking into a microphone. The closer your mouth gets to the microphone, the more the low frequencies are accentuated. One solution is to step away from the microphone, but there might be situations where that is not practical. Therefore, we need other solutions to the problem.

Even though I did state earlier that you should avoid trying to fix issues in the DAW, the proximity effect is easiest solved by placing a high-pass filter on the recording to lower the bass. Some microphones include a cutoff switch that will engage a high-pass filter at 75 Hz to help reduce the proximity effect. The effect is most pronounced with cardioid pattern microphones. Omnidirectional microphones are less prone to the effect.

Low-Frequency Rumble

Vibrations on the floor can travel up the microphone stand and cause the microphone clipped to the stand to vibrate. These vibrations will translate to a rumble in the audio signal. Using a high-pass filter in the DAW or on the microphone may help alleviate this issue. The best method is to use a microphone shock mount. The shock mount suspends the microphone in the center of a ring with elastics. Any vibrations from the microphone stand dissipate across the elastics, thereby reducing the rumble effect. If you are placing the microphone on a hardwood floor, then placing a rug or mat underneath the stand will help.

Plosives

Plosives are the bursts of air that occur when saying or singing consonants like the letters B, P, and T. The plosives are useful in determining which consonants are being used in a sentence. However, they can present problems when using microphones. If you say "perfectly placed petunias" into microphone you may hear the letter P in each word as small low-frequency explosions on your recordings, which are distracting. The easiest way to prevent plosives is to use a pop filter in front of the microphone. A pop filter is a nylon mesh stretched between a ring and placed in front of a microphone. The mesh helps diffuse plosives and reduces the chances of them being picked up from the microphone.

Pop filters are extremely useful, and you should include one in your home studio if you plan on recording voices speaking or singing.

Microphone Placement Techniques

There are other textbooks that go into detail regarding microphone placement. Rather than duplicating that information, I will offer you some practical guidelines to get you started.

Microphones can be placed close or distant. Close placement captures the sound of the instrument or voice directly with little of the room. The microphone is usually between one and three feet from the source. Close placement is useful for capturing the character and tone of the instrument or voice. Attacks are very clear, and you can hear the details of the instrument and voice. If the microphone is too close, you may end up capturing details like breathing or instrument noise that is not flattering to the sound you are trying to achieve. If you are capturing too much detail, try changing the axis of the microphone to reduce the amount of direct sound. Listen to the results carefully and adjust until you find a good balance.

Distant placement is useful when you want to capture the sound of the voice or instrument in the room. Distant placement requires you to place the microphone at least four feet away from the source. The sound of the source and the environment is captured. The result is a light, open sound in comparison to close placement. Distant placement is useful when you are recording a group of musicians and want to capture the blended sound of all the musicians. You must be careful not to place the microphones too far away or else the sound may become muddy or unfocused.

If you have enough microphones and inputs, then a combination of close and distant placed microphones allows you to capture the instruments and rooms. You can then adjust the balance between the two sounds when you mix the tracks in your DAW. This is useful if there are specific instruments or voices you want to accentuate at different times.

Stereo Placement

Stereo placement involves two microphones spaced apart to create the illusion of a stereo image. With distant placement, a pair of microphones can create depth and distance between the left and right sides of the room. You can create a stereo illusion of a piano by placing one microphone near the low strings and another one near the high strings. When using multiple microphones, you must be aware of their placement to avoid phase problems.

Phase determines the starting point of a waveform, whether it is at 0, 90, 180, or any degree amount. Two waveforms starting at the same point are in phase. Two waveforms starting at different points are out of phase. Waveforms that are out of phase can cause frequency cancellation and remove the depth and clarity of the sound. Any time you use two microphones to capture a source, you create the possibility of phase issues. If two versions of the same sound arrive at each microphone within milliseconds of each other and then are played back together, you have the possibility of phase cancellation. Here are a couple of ways to reduce the chances of phase cancellation.

The 3:1 Rule

The 3:1 rule states that for every unit of distance between the first microphone and its sound source, the distance of the second microphone should be at least three times as great. If using two microphones on a piano and the first microphone is positioned one foot above the strings, the second microphone should be placed at least three feet away from the first and be one foot above the strings. If you are recording a vocal group in a room and the microphones are six feet above the musicians, the second microphone should be 18 feet from the first while remaining six feet above the choir.

Mono Monitoring

When using two microphones on a source, we are always tempted to pan the track for each microphone all the way to the left and the right. The sound in our headphones sounds large and wide, but it is very difficult to hear phase issues in stereo. Once you achieve the desired sound, place each track in the center and listen. If you hear unnatural volume and frequency fluctuations with some frequencies reduced and others being boosted, then you have a phase issue. Adjust the placement of your microphones and check the results. The tone and character of the sound should not change with listening to the mix in stereo or mono.

Microphone Placement Suggestions

Below is a list of suggested placements for various instruments. This list is not comprehensive but will serve as a starting point. Remember, use your ears. If the sound is not to your liking, adjust the placement until you achieve the results you want.

Voice
Three to eight inches from the mouth
Adjust the axis of microphone up or down to change the tone

Acoustic guitar
Six to twelve inches from the sound hole
A second microphone can be placed the same distance at the bridge for more attack
A second microphone can be placed the same distance at the neck for finger and fret noise

Guitar or bass amplifier
One to twelve inches from the center of the speaker cone
One to twelve inches from the edge of the cone for a different tone
Adjust the axis left or right to change the tone

Upright piano
One to six inches from the soundboard, which is at the back of the piano
A second microphone can be added for a stereo sound

Grand piano
One to six inches from the strings with the lid open
The height of the lid will affect the sound you capture
A second microphone can be added for a stereo sound

Saxophone, trumpet, or other brass
One to six inches from the bell
Adjust the axis up or down to change the tone

Violin, viola, cello, double bass
One to eight inches from the sound holes
Closer to the bridge for more attack and grit
Closer to the neck for a smoother sound

Kick drum
If there is a hole in the drumhead, then start inside about one inch and pull out until the desired sound is achieved
Adjust the axis to change the tone
If there is no hole, start about an inch away toward the lower edge
Adjust the axis to change the tone

Snare drum and toms
One inch away from the top head toward the edge away from the player

Hi Hat
Place microphone two inches above the edge of the cymbal
Adjust so that the hats are open it does not strike the microphone
Adjust the axis to change the tone

This list is a starting point for microphone placement. Listen to your results and adjust. You do not need an expensive microphone to achieve a good sound. Microphone placement can make an enormous difference.

Monitors

The professional term for speakers is monitors, but sometimes people confuse the term with computer displays. You may have noticed that I have not used the term monitor until now and will continue to do so for the remainder of the chapter.

Monitors, or speakers, are an important part of your set up. Numerous variables such as room size, acoustics, and placement affect how monitors sound. Purchasing monitors is a subjective venture and opinions differ between individuals. There is no easy answer to the question "What is the best monitor to buy?" To be honest, the best monitors to buy are the best ones you can afford. Let us begin by exploring the design and function of a monitor.

Monitor Design

Monitors, like microphones, are transducers, except that they work in the opposite direction of microphones; monitors convert electrical energy into acoustic energy. The design of a monitor is nearly identical to that of a dynamic microphone. Monitors have a cone attached to a voice coil. A magnet sits inside the coil allowing the cone to move back and forth freely. When current reaches the voice coil, it reacts with the magnet and electromagnetic induction causes the cone to vibrate. The rapid motion of the cone compressing the air molecules creates sound. The entire assembly of the cone, voice coil, and magnet is called a driver (Figure 7.14).

Drivers

Most studio monitors have at least two drivers of different sizes. The size of each driver is suited for a particular frequency range. Having multiple drivers allows manufacturers to tailor the driver to operate best in the frequency range. The different sizes help improve the response and clarity of sound of the overall monitor.

The woofer is the largest driver and is best suited for low frequencies ranging from 80 Hz to 500 Hz. Woofers range between eight and 12 inches in diameter. The size of the woofer requires a larger magnet and therefore more current to move the cone.

The midrange driver produces sounds in the mid-frequency range from 500 Hz up to 4,000 Hz. These drivers range between four and eight inches in diameter. Depending on the size of the overall monitor, the midrange driver might be the largest driver in the monitor.

The tweeter is responsible for producing sounds above 4,000 Hz. The tweeter ranges from one to four inches in diameter. There are several designs for tweeters, some of which do not use a conventional cone to produce sound.

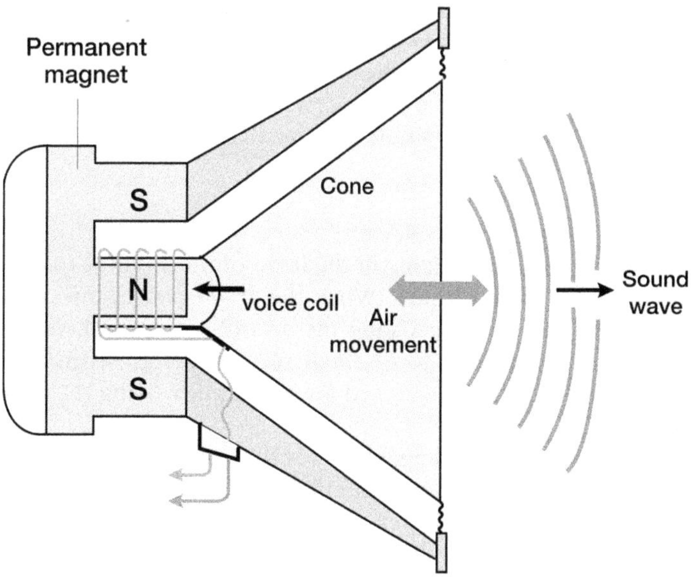

Figure 7.14 Image of interior of monitor

Monitors with three drivers are called three-way monitors. Monitors with two drivers are called two-way monitors. Three-way monitors are generally more expensive and larger since the cabinet must house three drivers. Home studio monitors are generally two-way and come in sizes small enough to size to place on your desk close to you.

Enclosures

All the speaker components are placed inside an enclosure. The enclosure is an integral component in speaker design and shapes the way a speaker sounds, particularly the low-end response.

Sealed Enclosures

Sealed enclosures do not allow air in or out of the enclosure, hence the name. The air pressure inside the enclosure reacts to the motion of the drivers. When the drivers push outward, the air pressure decreases, and when they push inward, the air pressure increases. Remember that air molecules are constantly trying to return to their original state. Like our balloon example in Chapter 2, the air in the enclosure will push against the pressure of the drivers when they push inward. This pressure acts like a spring and pushes the driver outward with greater force.

This action and reaction create a tighter, more precise sound production. This changes the way we hear the low frequencies. The bass response is not as loud as ported enclosures, but there is more force and more punch from the speaker. The resulting bass sound hits you in the chest and you feel the lower frequencies more than you hear them. However, to achieve a big and powerful bass sound, you need large drivers and powerful amplifiers. The result is large enclosures with woofers larger than ten inches.

Ported Enclosures

Ported enclosures have an opening in the front of the monitor to allow air to move in and out of the enclosure. When the driver pushes inward, it forces the air out of the enclosure, boosting the overall sound level, especially the low frequencies. This allows for a small speaker design with a big sound without the need for large drivers and amplifiers, increasing the efficiency of the monitors.

The main issue with ported enclosures is the precision of low frequencies. The ports allow for a bigger sound, but the sound is not as focused as a sealed enclosure. This can make low frequencies sound blurry, making it difficult to distinguish the sound of a kick drum against a bass guitar. Many of us will enjoy the bigger sound, but in time we may have difficulty adjusting low frequencies in our mixes.

Amplifiers

Monitors can be passive or active. Passive monitors require an external amplifier to power the drivers in a monitor. Active monitors have amplifiers built into the monitor cabinet. Nearly, all monitors designed for home studio use are active monitors.

Each driver in a monitor requires an amplifier to create the current to create the electromagnetic field. Woofers have larger magnets and require additional power from an amplifier. Midrange drivers and tweeters require less power. Budget monitors will use one amplifier to power all drivers. More expensive models will use dedicated amplifiers for the individual drivers. Dedicated amplifiers allow manufacturers to tailor and tune the amplifiers for each driver. The decision depends on the manufacturer and the cost of the monitor. Refer to the specifications of the monitors to determine the number of amplifiers included in the speaker.

Subwoofers

Subwoofers are monitors specifically designed to enhance low frequencies below 200 Hz. These are ported enclosures to allow the buildup of low frequencies. The larger the enclosure of a subwoofer the greater the efficiency and louder the output. The low frequencies will ring out longer, giving the impression of a bigger and fuller sound. Subwoofers make the low end more audible, but not more precise. Remember that these are ported enclosures, which are not known for their precise low-frequency reproduction.

Low frequencies are non-directional, meaning that it is difficult to determine the direction that sound is coming from. This means that you can place a subwoofer anywhere in a room; it does not have be to in the center underneath your desk. Placement against walls is good as well, since the walls will increase the loudness perception and help distribute the sound.

Subwoofers are necessary for surround sound applications in films and games. The enhanced low end adds to the audience experience needed for these types of media. Subwoofers are also effective for live concerts, where it is important to have low frequencies heard by all regardless of where they are sitting. With regards to a home studio in a small room, a subwoofer might make you feel good, but can create an artificial balance of your mix. Your music may sound great with the subwoofer engaged, but how will it sound on a system that does not have a subwoofer? Take the funds you are going to spend on a subwoofer and purchase higher-quality monitors.

Monitor Size and Placement

Many feel that purchasing larger monitors is better because they produce a bigger sound. It is important to understand that monitor placement and room design play a more important part in how a monitor sounds. The size of your room dictates the placement of your monitors, which in turn dictates the size of monitors you purchase. The larger the monitors, the more space you need between them for the sound to be optimal. A practical measurement to use is you need one foot of space between the monitors for every inch of the monitor size. A five-inch monitor requires at least five feet of space between them. Eight-inch speakers require eight feet of space.

For optimum listening, monitors must be placed so that they form an equilateral triangle between each other and you. Your five-inch monitors that are five feet apart from each other also need to be five feet away from you. Your eight-inch monitors need to be eight feet away from you. As you can see, large monitors require large rooms for the sound to be optimal. The further the monitors are from your ears, the more you need to consider the acoustics of your room you are in. Large monitors are suitable for far-field monitoring where the sound of the speakers interacts with the acoustic space of your studio to create an ideal listening environment (Figure 7.15).

The vertical height of the monitors should be high enough so that their tweeters are at ear height (Figure 7.16).

Another reference for determining monitor size is if your room is 15 × 15', then the monitors should be four to five inches in size. These monitors are designed for near-field monitoring. Near-field monitors need to be placed close enough so that you hear the direct sound from the speakers and not the room. Having the monitors close to you reduces the audible effect of the acoustic in your room. Most of us are not working in acoustically treated rooms, so it is best to leave the room out of the listening environment.

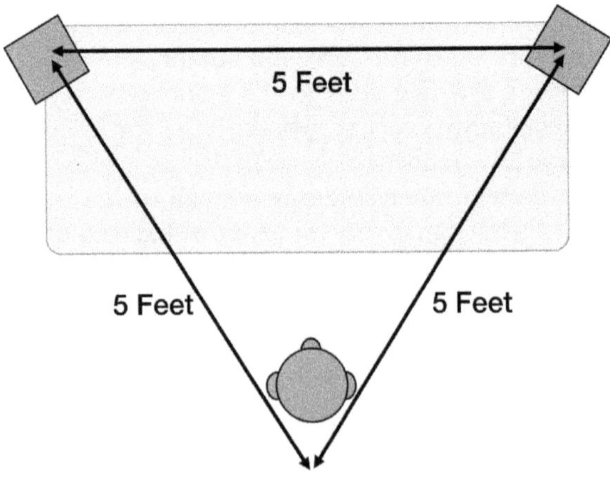

Figure 7.15 Image of equilateral triangle monitor placement

Figure 7.16 Image of monitor height placement

Choosing Monitors

Purchasing monitors is subjective and pricing varies. Bring music you are familiar with as some stores will allow you to listen to your music. This can be helpful when selecting monitors. Keep in mind that a pair of monitors may sound great in the store but may respond differently in your room. Remember that the environment that you work in will influence the sound of the monitors. You may find that monitors have too much or not enough low frequencies in your room. Save the box and your receipt in case you need to return monitors and try another pair.

Headphones

Headphones are helpful if you have a less than ideal listening environment. Perhaps you live in a busy area with traffic noise, or you are in a quiet area where you cannot listen to music past a certain volume. In these areas, headphones allow you to keep working without being disturbed by the environment or disturbing your neighbors. However, creating accurate recordings and mixes can be problematic with headphones.

We hear sounds all around us in a 360-degree circle. When we listen to music through monitors, we are hearing the sound characteristics of the monitors along with the room we are in. Our perception of the direction of sound is determined by both ears. Sound does not always arrive to our ears at the same time, there might be a slight delay between the two ears. These delays help us determine the direction of the sound. The delays also help our brains assemble which sounds are in the center. The vocals on a recording sound in the center of our hearing field because both ears are receiving the sound at nearly the same time.

With headphones, sound arrives at both ears at the same time without any acoustic interference. We do not hear how the sound interacts with the room because we are not hearing the room. Headphones also prevent us from hearing sound in a 360-degree circle around our head. We can perceive the direction of sound based on how the sounds are panned in a stereo field. But the volume of the sounds in the center is not always accurate because our ears do not have the benefit of the acoustic environment. This means that vocals in the center may sound louder or quieter in headphones than they do on speakers. Whenever you mix music in headphones, you should always listen to the mix on speakers to determine if the information in the center is at the correct volume.

Headphone Types

Headphones vary by design, sound, and cost. Like monitors, choosing headphones is subjective. There are two basic types of headphones, closed-back and open-back. Each of these types has its advantages and disadvantages. Let us look at each type in more detail.

Closed-back designs block out outside noise. The back of each cup is sealed, limiting the amount of external noise you hear. These headphones will either sit directly on top of the ear or cover the ear entirely. The goal of these headphones is to isolate the listener from the outside environment. The advantage of this design is that you can set your listening levels much lower because you are not trying to block out the external noise. The headphones stay in place because the headband exerts pressure to press the cups into your ears. Some users find these headphones uncomfortable because of the pressure on the sides of your head. Those who wear glasses might find the pressure on their frames uncomfortable as well.

Open-back headphones allow you to hear the environment around you. You will find yourself using higher listening levels higher to block out external noise. These headphones are lighter in design and therefore much more comfortable. They are designed to fit snuggly, but with very little pressure on the sides of your head. These headphones can be worn for long periods of time. Many users prefer these types of headphones because they do not feel disconnected from the outside world. These headphones can also sit directly over the ear or on top of the ear.

Choosing Headphones

Headphones vary greatly in price and like monitors, preference is subjective. You benefit from visiting a store and listening to different models for comparison. Bring music that you are familiar with so that you have a reference point for listening. Professional headphones can sound different than consumer headphones. Professional headphones have a more neutral sound, and the frequency response is flatter. If you are used to headphones that enhance lower frequencies, then you might find professional headphones dull at first. Your ears will adjust to hearing music without enhancements. Once you find a pair of headphones you like and have grown accustomed to them purchase a second pair so that you have a backup.

Summary

In this chapter, we looked at the essential audio hardware needed for your home studio. Our goal is to discuss the essential equipment you need to start creating music. We did not cover other hardware such as external preamplifiers or effects units. Those of you interested in electronic music, we will look at synthesizers and keyboard controllers in the upcoming chapters. Even if you are looking at entry-level equipment, the cost of each item adds up quickly, so you need to be strategic with your purchasing decisions. I have given you suggestions on how to determine your needs and working within your budget. As your experience grows and needs grow, you want better equipment and find yourself upgrading some of your hardware. For now, figure out what you need to get started and build from there.

Activity 7

Audio Hardware Recommendations

Like the computer recommendations in Activity 4, this list will help you determine which hardware you need to get started, depending on your production needs. You may already have some hardware and this list can help you determine if you need to upgrade equipment. Depending on your budget, these recommendations can help you determine the price difference between what you need to get started and what you need to do everything you want. Some of you may already have enough equipment to get started; you can plan for upgrades later. All these recommendations assume you are purchasing entry-level equipment. In Chapter 11, we will look at upgrade strategies and how to plan for those occasions. I am not recommending specific brands or models because the music manufacturing market changes often and models are often discontinued after a year.

Recommendations

Audio Interface

2 inputs and 2 outputs if working alone with occasional guest
4 inputs and 4 outputs if working a few musicians
8 inputs and 8 outputs if recording drums or an entire band
USB connectivity
Thunderbolt may add more inputs and outputs but at a cost

Microphones

A single condenser microphone if working alone and recording voice or instruments
A single dynamic microphone if performing live
2 condenser microphones if recording instruments and voice
Multiple microphones to record drums

Headphones

Open back headphones are the most comfortable
Ideal if you will be using them a lot
Closed back models are helpful if working in loud environments
Avoid models that offer audio enhancements
Wireless models can offer lower fidelity

Speakers

Smaller is better as you can place them on your desk close to you
It is better to spend money on a better set of smaller speakers than a larger set
All entry-level speakers will be ported so the bass response will vary
If your desk is close to the wall, choose a model that lets you adjust the low- and high-end frequencies
Models with integrated Bluetooth are handy, but not necessary

MIDI Controller

Experienced keyboard players will want at least 61 full-size keys
Composers and songwriters may prefer 88 full-size keys
Most controllers offer eight pads for programming percussion parts
If you want extensive control of your DAW application, choose a model that integrates easily
Mini controllers are ideal for portable systems
Some controllers are designed for specific DAW applications only

Conclusion

Read reviews of everything you are considering for purchase. Visit the manufacturer's website and download the manuals. The manuals will tell you how the hardware operates and its features and specifications. If you have further questions, contact the manufacturer directly. Many of them are very good at answering questions and extremely helpful. Take your time when purchasing audio hardware.

Terms

ADAT An eight-track digital audio recorder developed by Alesis that could record eight tracks at a sample rate of 48 kHz and a bit depth of 24 on S-VHS tapes.

Balanced Signal An audio signal traveling over a TRS or XLR cable.

Direct Injection A hardware device that converts high-impedance signals to low impedance

Impedance The amount of resistance present in the electrical current in the cable.

Instrument Level A signal level with high impedance produced by instruments such as guitars and basses.

Latency The time it takes the computer to process an input signal and then send it back out.

Line Level A low impedance high-level audio signal produced by most audio equipment and synthesizers.

Microphone Level A low impedance low-level audio signal produced by microphones.

Preamplifier An amplifier designed to significantly increase the voltage of an incoming microphone-level signal.

Sony/Philip Digital Interface (S/PDIF) Consumer-level digital format developed by Sony and Philips.

Unbalanced Signal An audio signal traveling over a TS cable.

Chapter 8

MIDI

History of MIDI

MIDI stands for Musical Instrument Digital Interface. MIDI is a protocol that was developed to connect synthesizers together so that they could control each other. Imagine having two synthesizers, one with a piano sound, another with a string sound, and you want to layer the two sounds so that the piano and strings would play the same notes. Prior to MIDI, you would need to play both synthesizers at the same time to layer the sounds. MIDI allowed one synthesizer to control the other, meaning that you could play the piano synthesizer and trigger the notes on the synthesizer with strings.

MIDI is data and nothing else. Sound is not transmitted, only note information. When you trigger notes on the second synthesizer, you are only sending the notes you want the second synthesizer to play. The sounds created must come from the synthesizer. MIDI tells the other synthesizer what notes you want to play, how loud you want them to sound, and how long you want them to sound. Since MIDI is only transmitting and receiving data, it is fast and reliable.

In 1981, Dave Smith, the founder of Sequential Circuits, the synthesizer manufacturer that created the classic Prophet-5 synthesizer, wanted a way to play two synthesizers at once. He created a protocol that he called Universal Synthesizer Interface (USI). USI used a microprocessor to convert the voltage pitch information from one synthesizer into serial, or numerical, data. This data was sent over a cable to another USI synthesizer, allowing it to be controlled by the first. Dave Smith's attempt was successful, and its development attracted the attention of Ikutaro Kakehashi, the president of Roland. The two companies worked together for the remainder of the year to create a protocol that could be used by all synthesizers.

At the 1982 Audio Engineering Society (AES) conference, the two companies demonstrated the power of MIDI by having two synthesizers from two different manufacturers controlling each other via MIDI. In 1983, the MIDI specification was finalized at version 1.0 and approved by the newly formed MIDI Manufacturers Association. In 2013, Ikutaro Kakehashi and

DOI: 10.4324/9781003345138-9

Dave Smith were each awarded a Technical Grammy for the development of MIDI. The MIDI 1.0 version was so successful and stable that version 2.0 did not go into development until the year 2020.

When working with a digital audio workstation (DAW) MIDI is transparent to you, but it is still present any time you use a virtual instrument controlled with an external MIDI controller. Even though your controller is connected over Universal Serial Bus (USB), it is transmitting MIDI messages. Every note on the virtual instrument is assigned a MIDI note number. When you record a virtual instrument, you are recording MIDI data. MIDI is flexible because you can quickly change instruments and still retain the same notes. You can change the tempo without affecting the sound. You can even transpose the pitches to a different key.

MIDI Basics

As I stated earlier, MIDI is a communications protocol that sends and receives data. The data packets are extremely small, eight bits in size, which is the equivalent of one typed letter in this book. MIDI can only communicate with devices that are MIDI compatible. MIDI can transmit patch and note messages. This means it can tell the receiving unit to use sound 32 and play middle C. Keep in mind that the transmitting synthesizer does not necessarily know what sound 32 is on the receiving synthesizer. When connecting a sequencer to a drum machine, MIDI can transmit start, stop, and tempo data, which tells the drum machine when to start playing, how fast to play, and when to stop.

MIDI does have limitations. MIDI is one-way communication. This means if you have a synthesizer that needs to send and receive information, then you need two MIDI cables. Each MIDI connection can send or receive only 16 channels of information. Each channel is unique and sends independent information to each channel. The range of MIDI is limited to 128 “degrees” of resolution. A MIDI device can send either a range of data from 0 to 127 or 1 to 128. This means that if you are using MIDI to control the volume of an instrument, the loudest at the instrument will be a value of 127 and the lowest would be 0, which is no sound.

MIDI Channels

MIDI can transmit and receive information on 16 independent channels. Each channel can send and receive unique information. Receiving MIDI devices must have the same channel number as the transmitting device. If the first device is on MIDI channel two and the second device is on MIDI channel three, the second device will not receive any of the data transmitted from the first device. MIDI devices generally have a method of setting the MIDI channel, although units such as drum machines are often set to channel 10. We will find out why drum machines use channel 10 later in this chapter.

To make things easier, many MIDI devices have global settings that allow you to specify whether the device will receive and transmit on a specific channel or multiple channels. These settings are known as MIDI Modes. When the MIDI mode is set to OMNI ON, then the device will send and receive on all MIDI channels. A receiving device set to OMNI ON will receive all data from the transmitting device, regardless of the MIDI channel. If you want a device to send and receive on a specific MIDI channel only, then you can set the MIDI mode to OMNI OFF.

Legacy MIDI devices can be set to play multiple notes at the same time such as chords, polyphonic, or one note at a time, monophonic. A device in POLY mode will play polyphonically, meaning that it can play multiple lines and chords. A device in MONO mode forces the instrument to play monophonically, meaning that it can only play one note at a time.

MIDI Cables and Ports

MIDI transmits information over a five-pin male DIN connector. The DIN connector is a standardized electrical connector developed in the early 1970s. Rather than creating a new cable, the MIDI specification used an existing connector. MIDI only uses three of the five pins on the connector. The maximum length of a MIDI cable is limited to 50 feet. The MIDI cable is not balanced or shielded. This means it can be affected by radio magnetic and electromagnetic interference. A single MIDI cable can transmit over 1,000 messages per second.

MIDI specifies that devices can have up to three MIDI ports. The three ports are IN, OUT, and THRU. The IN port receives up to 16 channels of MIDI data. The OUT port transmits up to 16 channels of MIDI data. The THRU port passes information received via the IN port so that you can daisy-chain multiple devices. This port functions as a pass-through for information; it will not transmit data from the connected synthesizer. If you have two synthesizers and want to send data from one to the second, you connect the OUT port from the first unit to the IN port of the second. If you connect the THRU port from the first unit to the IN port of the second, the first unit will not control the second unit (Figure 8.1).

The THRU port is used when you have two synthesizers that you want controlled by a single unit. In this scenario, you connect the OUT from the first unit to the IN of the second. Then, you connect the THRU from the second unit to the IN of the third. This way, the second and third units both receive data from the first. You can theoretically daisy-chain unlimited units, but you might experience delays after the third connection. If you need connect multiple devices to a single device, then you need a MIDI THRU box, which has a single MIDI IN port and multiple THRU ports. Each secondary

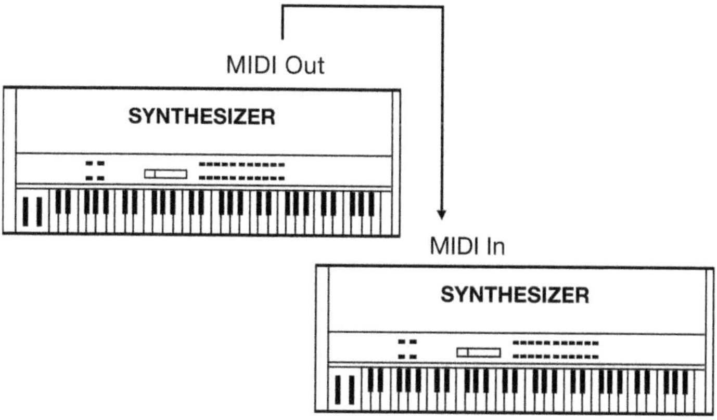

Figure 8.1 Two synthesizers connected from MIDI OUT to IN

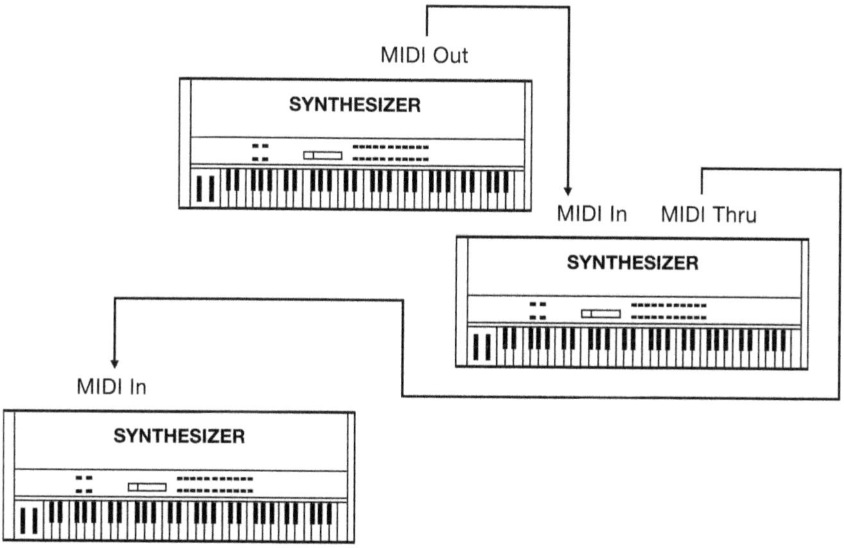

Figure 8.2 Three synthesizers connected from MIDI OUT to IN, then THRU to OUT

device connects to a single THRU port on the box. Keep in mind that each secondary unit must be set to the same MIDI channel as the transmitting unit (Figure 8.2).

A MIDI THRU box simplifies the connections and helps with organization (Figure 8.3).

Modern synthesizers still have MIDI ports to connect to legacy devices, but most will have a USB to connect to your computer. Even though it is

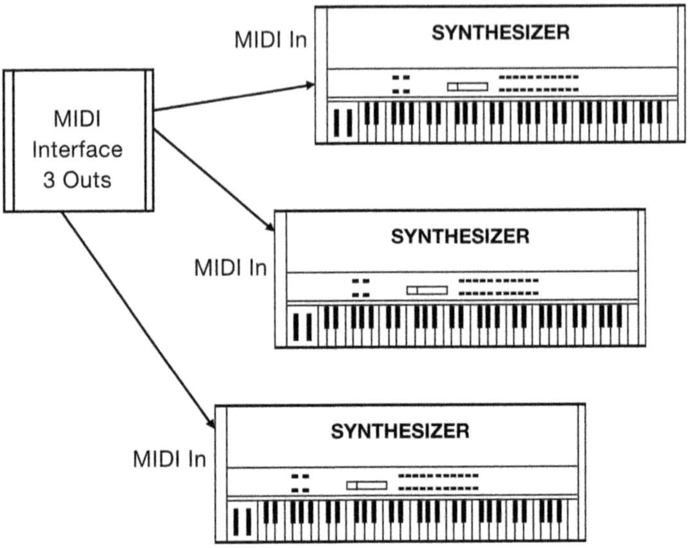

Figure 8.3 Three synthesizers connected using a MIDI THRU box

a USB connection, your computer is transmitting and receiving MIDI data over that connection. If you have a legacy device that only has MIDI ports and you want to connect it to your computer, then you must use a MIDI interface. If the audio interface connected to your computer has MIDI ports, then you connect the device using those ports. If your audio interface does not have MIDI ports, then you need a MIDI interface.

MIDI Interfaces

If you wish to connect legacy MIDI devices that only have MIDI ports to your computer, then you need a MIDI interface. A MIDI interface allows your computer to send and receive MIDI data. MIDI interfaces connect to your computer using USB and many or class-compliant meaning they do not need drivers installed. Some audio interfaces offer a single MIDI IN and OUT connection, which may be enough for many of you. You can even daisy-chain multiple devices if you wish. However, for the most flexibility when using multiple devices, it is best to use a multi-port MIDI interface.

A multi-port MIDI interface allows you to connect multiple MIDI devices independently, meaning that each device can send and receive data over all 16 MIDI channels. This means if you have multiple multi-timbral synthesizers, you can use all 16 channels on each device. Multi-port MIDI interfaces help avoid latency, which can occur when daisy-chaining multiple devices.

MIDI Messages

MIDI transmits two types of data: Channel Messages and System Messages. Channel messages are specific to each MIDI channel. This means that each MIDI channel can receive different types of channel messages. These messages are performance oriented, meaning that the information sent and received all pertain to notes. System messages are global and affect all MIDI channels. Messages such as tuning are global since it affects the tuning of the entire synthesizer.

Note Messages

There are three categories of channel messages: note, expression, and program. Note messages pertain specifically to note information. When you press a note on a MIDI keyboard, you are sending a Note On message. When you release the note on a MIDI keyboard, you are sending a Note Off message. The note message must also include the MIDI channel number so that the note data is sent to the correct synthesizer. The note number specifies which note is being played. The velocity data specifies how hard the key was pressed in the range of 0–127.

MIDI assigns a unique number ranging from 1 to 128 to each note. It might seem that 128 notes are limited, but keep in mind that a piano only produces 88 notes, and that most music produced fits within the range of the piano. Middle C, which is the center pitch on the piano is assigned note number 60. The C# above that note is note number 61 and the B middle C is note number 59. The lowest note on the piano is note number 21 and the highest note is note number 109. A five-octave keyboard on a synthesizer or MIDI control transmits notes from 36 to 96. You can transpose the keyboard up and down to produce notes below and above that range. Figure 8.4 provides you with a visual reference of MIDI note numbers against the keyboard.

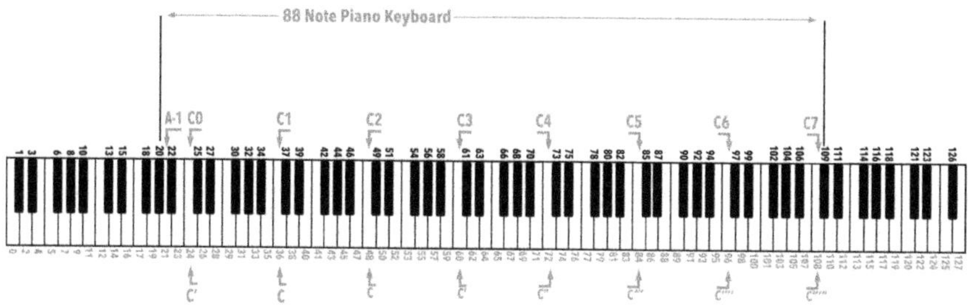

Figure 8.4 MIDI note numbers referenced against a piano keyboard

Expression Messages

Expression messages affect the note after the Note On message is transmitted. Expression messages affect the way the note sounds after it is played. There are several types of expression messages. Control change (CC) messages can affect the volume (CC 7), stereo placement, or panning (CC 10), modulation (CC 1), and other variables of the note. There are 128 possible CC messages, but not all of the messages are assigned to a function. The unassigned message can be customized by the user. One of the most common control messages is CC 64, which is sustain. When you attach a sustain pedal to your MIDI keyboard you press the pedal down so that the notes you just played continue to sound even after you release your fingers from the keyboard. You can use the modulation wheel on your keyboard, which is assigned to CC 1, to add vibrato or variation to the notes you are playing. Table 8.1 shows the most common CC messages you will encounter.

Channel Pressure

Channel Pressure messages are available on newer synthesizers. Channel pressure is a feature built into the keys of a synthesizer that allows you to press down on the key further after you play the note. Adding pressure to the keys after you play the notes allows you to affect the sound. The most common use of channel pressure is to add vibrato or other modulation to the sound. Channel pressure affects all notes being played at the same time. If you play a four-note chord and apply pressure to only one key, all four notes will be affected by that single key. Polyphonic key pressure allows each key to send individual pressure, meaning that if you play a chord and pressure to a single key, only that note will be affected. Keep in mind that not all MIDI devices can receive or transmit channel or polyphonic pressure. Whenever you send a CC message that a keyboard cannot respond to, it simply ignores the message.

Table 8.1 Common MIDI CCI numbers and descriptions

CC Number	*Description*
1	Modulation
7	Volume
10	Pan
11	Expression
64	Sustain
91	Reverb amount
93	Chorus amount
123	All notes off

Pitch Bend

Pitch Bend messages allow you to bend the pitch of the note up or down. Pitch bend is different from other MIDI messages because its range is from -8,192 to +8,191 instead of 0 to 127, which is odd. Pitch Bend messages consist of two values: course bend and fine bend. Course bend corresponds to MIDI note numbers and ranges from 0 to 127. A coarse bend message of 12 would bend the note up 12 half steps, or one octave. Fine bend messages range from 0 to 127 and correspond to the values between each note number. This means that pitch bend transmits two sets of messages. There are 128 possible notes and another 128 steps between each note meaning we have 128 times 128 possibilities or 16,384 values. A pitch bend wheel can bend notes up or down, which means half of these values must sit below zero and the other half must sit above zero. This means we end up with a range between -8,192 and +8,191.

Program Messages

Program messages are also called patch change messages because these messages tell the synthesizer which patch or sound to load. Legacy synthesizers shipped with 128 different sounds, which directly corresponded with the range of MIDI. As synthesizers developed the number of stored patches increased. Fortunately, the MIDI specification allowed for bank changes. This allowed a synthesizer to have up to 128 banks with 128 sounds in each bank which means you can have up to 16,384 patches on a single synthesizer. Program change messages first send the bank number followed by the patch number. Synthesizers with a single bank of 128 patches would ignore the bank message.

System Messages

MIDI System Messages are messages that affect the entire system on all MIDI channels. There are two types of system messages: common and real-time messages. Common messages were important when using hardware sequencers. These messages would tell the sequencer which song to load and where to start in the song. You could also use these messages to tune all your synthesizers. System exclusive messages were used to reset a synthesizer to a default state or load custom settings. System exclusive messages were unique to each manufacturer and device. System exclusive message designed for one synthesizer would only work on that model. If you sent those messages to another synthesizer, it would ignore the information.

Real-time messages are also related to hardware sequencers. These messages told a sequencer to start or stop playback. These messages also provided a timing clock for other sequencers or drum machines to follow. One common workflow was to program the drum patterns on a drum machine and the remainder of the music on a sequencer. The two devices were connected over MIDI and the sequencer would send start and stop messages to the drum machine and also provide a reference so that both units remained synchronized while playing.

A lot of these features seem trivial to those of us working the DAWs, however, at the time, the ability to synchronize devices was extremely important. I should mention that several manufacturers are producing hardware sequencers for users who want to work with physical instruments rather than virtual ones on a computer. These new hardware sequencers still use real-time messages.

General MIDI

By the late 1980s, MIDI was present in every synthesizer, drum machine, and sequencer. MIDI capabilities were added to computer sound cards, and some added a simple synthesizer to play back MIDI files without needing a synthesizer or MIDI interface. The development of multi-timbral synthesizers meant that you could purchase a single device that could play back different sounds at the same time, each on its own MIDI channel. The only issue was that each manufacturer had a different implementation of features. For example, a MIDI rhythm track created on one drum machine would not map correctly with a different drum machine. The first drum machine would assign the kick drum to note 60 and the snare to note 61 while the second drum machine assigned the kick drum to 48 and the snare to 50. The acoustic piano patch was number 11 on my synthesizer but number 22 on a different one.

The MIDI Manufacturers Association recognized the problem and decided that there needed to be a standard specification for MIDI implementation for synthesizers, computer sound cards, and consumer devices. The hope was to create a MIDI standard that all manufacturers could follow that would allow better interoperability between devices. The standard would also affect the type of MIDI file created so that compatible devices would read all the data stored within the file. This standard became known as General MIDI or GM. Devices that met the General MIDI standard would have the letters GM or General MIDI added to their front panels.

The GM standard listed a series of specifications for MIDI devices:

- Standard patch map of 128 sounds
- 16 sound banks with eight sounds per bank
- Minimum of 24-note polyphony
- Minimum of 16-part multi-timbral capability
- Minimum of 16 MIDI channels
- Standard percussion mapping
- Standard for MIDI files

Patch Maps

The standardized sound sets would ensure consistent playback between manufacturers. For example, if I create a MIDI file and choose an electric piano to playback on channel 1 on a Roland synthesizer, playing back that same file on a Yamaha synthesizer would still use an electric piano on channel 1. The actual sound of the electric piano would vary from manufacturer to manufacturer, but the patch would be an electric piano. Standard patches allowed users to share files without having to re-map sounds.

General MIDI patch maps use 16 bank families, each with eight sounds within the family. The first bank contained eight different piano sounds such as acoustic and electric pianos. The banks ensured that the correct sounds were loaded when a MIDI file was shared. Table 8.2 shows the patch mapping for GM.

Multi-timbral Standards

Standardized patch mapping allowed users to share MIDI files between different multi-timbral synthesizers. A multi-timbral synthesizer can produce multiple sounds at the same time, each playing their own part. A multi-timbral synthesizer can play a drum track, bass, track, guitar track, and piano track all at the same time. Each track was assigned its own MIDI channel, and its own MIDI part.

To ensure greater compatibility, GM required devices to support at least 16 MIDI channels, each with its own sound, and at least 24-note polyphony. Each device needed to be able to play at least 24 notes at the same time. The allocation of notes or voices was dynamic. A drum part usually used one or two notes, bass guitar, one, electric guitar two to six notes, and a piano up to ten notes. Dynamic voice allocation meant that if the guitar was only using three voices and the piano needed more than ten voices the piano could use three voices not being used by the guitar. However, if the guitar needed six voices, it could remove them from the piano. The priority of the instrument was determined by the channel number. If the piano part required the greatest number of voices, you would set it to channel one.

Table 8.2 General MIDI bank families

Bank 1	Piano	Bank 9	Reed
Bank 2	Chromatic percussion	Bank 10	Pipe
Bank 3	Organ	Bank 11	Synth lead
Bank 4	Guitar	Bank 12	Synth pad
Bank 5	Bass	Bank 13	Synth FX
Bank 6	Strings	Bank 14	Ethnic
Bank 7	Ensemble	Bank 15	Percussion
Bank 8	Brass	Bank 16	Sound FX

A less important part could be channel seven. Later multi-timbral devices would see voice counts as high as 128.

Percussion Mapping

MIDI channel 10 was reserved for drum tracks. You could use channel 10 for other instruments, but the only way to access drum sounds was to set your device to channel 10. Drum sounds were assigned to specific note numbers so that they kick and snare, for example, would play correctly regardless of the device. The kick drum was set at note number 36 and that the snare drum at note number 38. The rest of the drum sounds were mapped similarly. The GM drum mapping is shown in Figure 8.5.

Standard MIDI File

The Standard MIDI File (SMF) defined how multitrack MIDI data is saved. The SMF stored note information and bank and patch information. This ensured that the correct sounds were loaded for each track. Two types of MIDI files were created to suit different markets. Type 0 files stored all MIDI channels with data onto a single track. Type 0 files were suitable for consumer devices and video games since these users only needed to read and playback the files. Type 1 files stored each MIDI channel with data

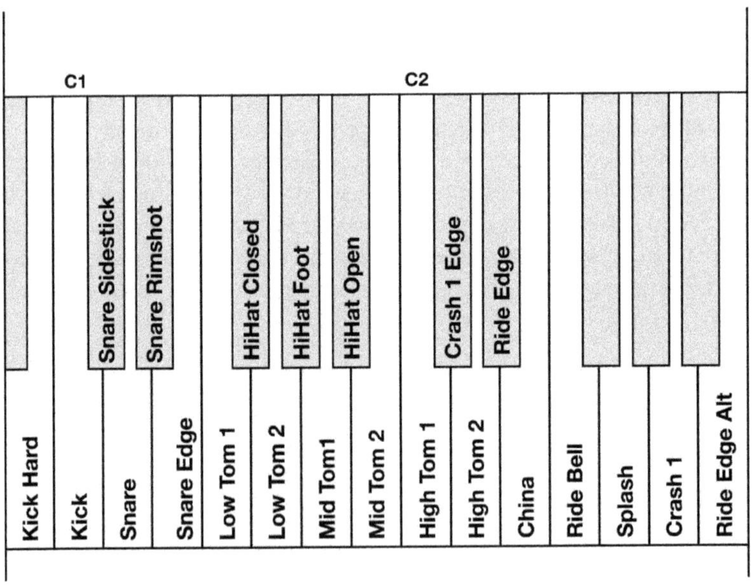

Figure 8.5 Drum MIDI note numbers referenced against a piano keyboard

on individual tracks. For example, a 16-channel MIDI file would open as 16 tracks, whereas a ten-channel file would only open ten tracks. Type 1 files were better suited for musicians who needed the ability to edit individual tracks.

The General MIDI standard is still used by all manufacturers because it offers a consistent template to ensure compatibility. There have been attempts to update the standard, but a single updated version was never established. Like MIDI 1.0, it appears that the initial GM standard continues to meet the needs of users today.

MIDI Controllers

As computers became more powerful, manufacturers developed virtual instruments, that is, synthesizers, that would load as plug-ins in your DAW. Consumers needed a device to control the synthesizers without needing to purchase a synthesizer. This need prompted the development of MIDI controllers. A MIDI controller is a device that allows you to send MIDI information via a MIDI cable or USB. A MIDI controller does not produce any sound; it only produces MIDI data.

MIDI controllers vary greatly in features and choosing one depends on your needs. Keyboard MIDI controllers have as few as 25 keys all the way up to 88 keys. The keys can either be full size, like that of a piano, or miniature keys which enable the device to be portable. The most common size is 61 full-size keys, which covers five octaves. To play all 128 MIDI notes, MIDI keyboard controllers can transpose up or down by octaves to increase their range. These controllers are velocity sensitive and there are different key beds to suit different players. Users accustomed to a piano will prefer weighted keys, which attempts to recreate the weight and feel of a piano keyboard. Others will prefer semi-weighted, which requires a lighter touch or even non-weighted, which is extremely light. If you are unsure which type you prefer, visit a store and try out different models; the differences are noticeable.

Some controllers offer just a keyboard with modulation and pitch bend wheels while others will have an assortment of eight knobs, eight sliders, eight buttons, and eight pads. Knobs and sliders transmit MIDI messages ranging from 0 to 127. Buttons are usually on/off switches to control playback of your DAW with stop, play, record, fast forward, and rewind functions. Pads are velocity sensitive and useful for playing drum parts. Some DAWS can automatically detect the controller you are using and automatically configure the unit to work with the DAW. Others allow for full customization with the DAW.

All these features and variations mean that the price range for MIDI controllers greatly varies. The best controller to purchase is the one that best meets your needs. If you have good keyboard skills, you might prefer semi-weighted or weighted full-size keys with at least 61 keys. Beat makers might prefer controllers with larger and more sensitive pads. Others may want a variety of knobs and sliders they can assign to their DAW to control virtual instruments and effects. Do your research and read user reviews before purchasing a controller. Take the time to try out different controllers so you can feel the differences in the key weights and features.

Synchronization

The last topic I want to talk about in this chapter is synchronization. At some point, you might need to connect your computer to an external sequencer, drum machine, or even another computer and play them back all at the same time. In this case, you need a way to ensure that when you press play on your machine that the other devices not only start up at the same time but also remain at the same measure and beat number at all times. You will need a way to synchronize all devices to a common clock.

A DAW uses an internal clock that is synchronized with the audio interface to ensure that every sample is processed at the correct moment and is aligned with each track. If you ever experience a slight delay between the moment you press play and hear sound, the DAW is processing all the samples in the buffer against the clock so that all tracks start at the same time. If your project has instrument tracks with virtual instruments, the DAW must also synchronize the MIDI information with the audio tracks.

Since your DAW is the primary machine in your workflow, you would need to establish it as the leader and the other devices as followers. To do this, you need to establish a clock base for all devices to follow. The easiest solution is to make your DAW the main clock for everyone else to follow. Now, you must determine what the clock base will be.

MIDI Beat Clock

The simplest clock is the MIDI beat clock, which is built into every DAW, hardware sequencer, and drum machines since it is part of the MIDI standard. MIDI beat clock, or MIDI clock is a clock signal sent at a rate of 24 pulses per quarter note. The clock is tempo dependent, which means the faster the tempo, the faster the rate of the pulses. MIDI clock uses bars and beats as its timebase. MIDI clock is simple to setup and instructions are included with most DAWs.

The primary issue with MIDI Beat Clock is that the lead system, your DAW, generates the clock for the other devices to follow but has no way of determining if those devices are keeping up with the clock. If one device falls behind or stops following the clock, the lead system has no idea of the failure. Chances are if you are synchronizing a short song, all your devices will remain locked, and any drift will be negligible. However, for longer projects, you probably need something more sophisticated.

MIDI Time Code

MIDI Time Code (MTC) uses absolute time as its timebase opposed to bars and beats. MTC uses hours, minutes, and seconds as a clock, which is more precise. This means that the clock is independent of the tempo. MTC is also part of the original MIDI specification, so it is built into all DAWs, sequencers, and drum machines. Because MTC uses real time as a clock, it is more accurate than the MIDI beat clock. However, since the clock information is sent along with other MIDI data, there is always a chance that the clock information may arrive late. Like MIDI beat clock, the lead machine cannot determine if the other devices are following in time or are falling behind. However, for most users, MTC is reliable and accurate enough for larger projects.

SMPTE Time Code

The Society of Motion picture and Television Engineers developed the SMPTE timecode to synchronize audio tracks to films. This allowed music and dialog to be added later with absolute accuracy and precision. SMPTE timecode uses absolute time for its timebase. However, since it was designed for film, SMPTE measures hours, minutes, seconds, and frames. The number of frames depends on the frame rate, measured in frames per second (fps) of the camera used to capture the film. The most common frame rates are 23.97 fps, 24 fps, and 30 fps.

Each device you want to synchronize must be capable of sending and receiving SMPTE. Your DAW can work with SMPTE, but your hardware sequencer and drum machine probably cannot. SMPTE also relies on a central synchronizer, which receives SMPTE data from all connected devices and ensures that all devices are always at the same hour, minute, second, and frame. If one device falls behind, the synchronizer will slow down the other devices until they all fall in line.

SMPTE is essential when working with film. It is extremely accurate and responsive. It is also useful if you have two DAWs, each playing the same music but with different parts, which need to be played back with absolute precision. However, for the home studio producer, SMPTE requires much more effort than the other two methods.

Workflows

If you are working with hardware sequencers and drum machines, the MIDI beat clock is the easiest to configure. For most users, its accuracy is sufficient for their workflow. If you need to synchronize two computers, MIDI time code is an excellent option. It is easier to configure than SMPTE and provides sufficient accuracy. I once used MTC to synchronize two computers, one handled the music from my DAW and the other had the movie I was working on. I did notice drift about 20 minutes into the film, the computer with the movie was slightly behind my DAW. However, all I had to do was stop both computers, start them again earlier in the film, and they remained synchronized for another 20 minutes. I saw this as a minor inconvenience since I was able to set up MTC in a few minutes.

SMPTE was designed for film, and if you end up working in film, you will learn how to use it in your workflow. However, for most of us, MIDI beat clock, and MTC will get the job done.

Summary

In this chapter, we looked at MIDI in detail and discovered that it is still important. Even though DAW software allows us to operate without thinking about MIDI channels, note numbers, and other specific details, understand that MIDI is working in the background. If you work with external synthesizers, especially legacy devices, then having a grasp on how MIDI operates will allow you to integrate that device into your workflow and DAW system. Some of you will be satisfied with virtual instruments and even working within the DAW. How you interact with instruments, real or virtual, will determine your needs and requirements for a MIDI controller. We also touched on synchronization, should you find yourself needing to synchronize multiple devices in real time. I would like to close this chapter by noting that even though MIDI was developed in 1982, it is still relevant even if it is still on version 1.0.

Activity 8

MIDI Editing

It is difficult to describe the process of MIDI editing so I thought we should spend a little time seeing how it is done. We will be using a free application called Aria Maestro. This is a very basic MIDI editor, but it allows us to get a feel of how MIDI editing is done within a DAW application. This application does not have any specific system requirements and will run on most Windows and Mac computers. You do not need a dedicated soundcard or MIDI controller to use the software.

Please follow this link to download the application for your operating system: https://ariamaestosa.github.io/ariamaestosa/docs/index.html. Your computer may question the security of this program when you try to launch it. Go ahead and allow the computer to run the application. It does not do anything to harm your computer. You will also need to download a MIDI file that I have provided for you on the companion website.

Once you have the application installed go ahead and run it. The main window will open and offer you three options. Use the third option which is Import a MIDI file. Click on the icon and load the Bach MIDI file you downloaded from the companion website.

The Piano Roll

The main window will display two lanes of information. Each lane shows all the MIDI notes that are in the file. You can press the play button at the top of the window to listen to the file. You may be familiar with this piece of music in may notice that there are several wrong notes in the file. We are going to correct these wrong notes.

If you look closely at the left edge of the first track, you will see a piano keyboard turned on its side. Notice how each piano key has a lane that extends all the way to the edge of the screen. The notes that are played in the track are blogs associated with each piano key and are spaced apart according to the note duration. This view is known as the piano roll. Every DAW application that allows you to edit MIDI will always display MIDI information in a piano roll. Some applications may call it by a different name, but they all function the same.

Tempo

The beauty of working with MIDI is that it is very easy to make corrections. We also have the option of altering the tempo of the performance. In the top menu area of the application, you will see that the tempo is currently set to 120. We can double-click on the number and change it to something else, for example, 90. Now when I press the play button, the piece will play slower. Now, we will edit the MIDI file to correct any wrong notes.

Note Editing

You will notice vertical lines spaced evenly across the tracks. These vertical lines represent the measure numbers, which are located just above the tracks. Our first wrong note occurs in measure three of the top part. If you look at each of the note blocks you will see the pitch name. The last note in measure three is F5. The note should be F-sharp 5. Select the note F5 with your mouse so that it turns green. Now, press the up arrow on your computer keyboard to move the notes up to F-sharp.

Anytime you edit a MIDI note, you will hear the pitch of the note that you edited. You have now edited your first MIDI note.

In measure seven, we have another incorrect note. The height of the track prevents us from seeing it, so you need to go to the far right edge and click on the down arrow to scroll to track down. You should now see F4 as the first note of measure seven. And then push the up arrow to move the note to F-sharp.

We need to fix a few notes in the lower part as well. The first note that we need to fix is in measure five of the lower part. The note at measure five is currently F#3. Select the note with your mouse and use the up arrow to move the note to A3. The next note to correct is in measure eight. The third note of the measure is currently A3. Select the note with your mouse and then use the up arrow to move the note to C4.

Conclusion

Now if we press the play button, we can hear the first eight measures of this piece correctly. There are several more incorrect notes in this file, but we are not going to fix them in this activity. I wanted to give you a brief introduction into the process of MIDI editing. If you are used to reading music, you may find this way of editing a little awkward at first. After some practice, however, you will find that editing MIDI is very efficient using the piano roll.

This concludes this activity. If you want, you can save the project or quit without saving.

Terms

Audio Engineering Society (AES) A professional organization made up of engineers, scientists, developers, and audio technicians who work together to evaluate and determine audio standards for professional audio.

Channel Message A MIDI message for a specific channel.

Channel Pressure A feature built into the keys of a synthesizer that allows you to press down on the key further after you play the note to alter the sound.

Control Change (CC) A MIDI message that can control a variety of MIDI functions. Many CC messages are standardized.

Expression Message Messages occurring after the Note On message that can be used to alter the sound.

General MIDI A MIDI standard for MIDI devices that required specific patch families and drum note mapping.

MIDI Beat Clock A simple synchronization clock that sends 24 pulses per quarter.

MIDI Interface A device attached to a computer that allows MIDI devices to communicate with the computer.

MIDI Time Code An absolute timebase system used to synchronize MIDI devices in real time.

Multi-Timbral Synthesizer A synthesizer capable of producing different sounds for each MIDI channel.

Program Message A MIDI message that changes the current sound patch of a synthesizer.

SMPTE Time Code An absolute timebase developed by the Society of Motion picture and Television Engineers to synchronize music and film.

Standard MIDI File A MIDI file that conforms to the General MIDI specification that can be shared with all MIDI devices.

System Message A MIDI message that affects the entire connected MIDI system.

Chapter 9

Synthesis and Sampling

Synthesis and Sampling

Synthesizers and samplers, whether physical or virtual, are important components for our home studios. Synthesizers offer a wide palette of sounds to add to our productions. Samplers allow us to include convincing pianos, orchestras, and choirs to our music as well. Most DAWs include a selection of virtual synthesizers and samplers. Therefore, we should take the time to learn about synthesizers and samplers so that we can include them in our workflows.

Synthesis

Synthesis is the process of creating sounds from scratch. A synthesizer is an instrument that creates sound waves using a variety of synthesis techniques. Synthesizers are useful for creating new sounds that cannot be produced by acoustic instruments. Synthesis begins with the creation of a waveform. The waveform can be generated using an electronic oscillator, a mathematical formula, or a computer algorithm. Once the waveform is created, the synthesizer can shape the sound by adding or removing frequencies, combining it with another waveform, or adding effects. Let us look at some of the more common synthesizer techniques.

Analog Synthesis

The first commercial synthesizer was created by Robert Moog and debuted in 1964. Moog's synthesizer design served as the template for analog synthesis as it is known today. Moog's synthesizer consisted of a series of modules, each with a different function. The modules were connected using 1/4-inch cables and followed a signal flow that is still used in contemporary analog synthesizers. Analog synthesizers use analog circuits to generate electronic sounds. One of the circuits in an oscillator, which produces a repetitive electronic signal. The repetitive electronic signal produces a waveform. The waveform

DOI: 10.4324/9781003345138-10

is the starting point of analog synthesis. The electronic waveform is usually one of the simple waveforms we discussed in Chapter 2: sine, square, and sawtooth.

The waveform is shaped by passing through a low-pass filter (LPF) or high-pass filter (HPF). The filters remove frequencies and shape the timbre or color of the sound. The resulting sound is then passed through an envelope generator (EG), which controls initial attack of the sound, the sustain volume, and how long it rings out. The resulting sound is fed into an amplifier, boosting the output level so that we can hear the result through our monitors or headphones.

Subtractive synthesis is a better way to describe an analog synthesizer. The filters that shape the sound subtract frequencies from the waveform, changing its timbre. Even though the term is more accurate, most people still refer to this method of synthesis as analog, as will we. An analog synthesizer is made up of three components: sound generators, sound processors, and controllers. Let us examine each of these components.

Sound Generators

Oscillators

A sound generator is an analog circuit that creates the waveform in analog synthesis. An analog synthesizer uses a voltage-controlled oscillator (VCO) to generate the waveform. The VCO is a circuit that creates a repetitive electronic signal. When an oscillator generates a sawtooth waveform, the sawtooth retains the same shape every single time you press the keyboard. The waveform does not change no matter how long you press the keyboard down. VCOs can exhibit unstable pitch when they are first turned on. Many early synthesizer manuals recommended waiting a few minutes after powering the synthesizer so that the pitch could stabilize as the voltage settled. To help with tuning, VCOs had a tuning control to fine tune the oscillator pitch should it go flat or sharp.

Each oscillator can produce a single note, meaning that it is monophonic. Early analog synthesizers could have up to three oscillators. The additional oscillators allowed you to layer waveforms to create more complex sounds. Each oscillator received the same pitch voltage from the keyboard so even though you have three oscillators, it was still a monophonic synthesizer. The number of notes you could play at once is described as voices. A synthesizer with eight-note polyphony can play eight voices at once. A monophonic synthesizer with three oscillators is described as having three oscillators per voice and only one voice.

The circuit in the oscillator determines the waveform. Analog synthesizer oscillators tend to generate five different waveforms: sine, square, sawtooth, triangle, and pulse. We learned about the first three waveforms and their timbre in Chapter 2. The triangle wave shares a similar harmonic pattern and the square, but the audio level of the harmonics decreases to a greater degree. The resulting sound is a cross between a square wave and a sine wave. It is harmonically more interesting than a sine wave, but not as colorful as a square wave (Figure 9.1).

The pulse is a variable waveform; the width of the waveform can be changed. In its center setting, the pulse wave is shaped like a square wave. You can narrow the pulse to create a waveform that looks like a narrow and tall rectangle. Widening the pulse creates a short and wide rectangular waveform. Altering the width of this waveform changes the number of harmonics in the waveform. The waveform can have a thin sound at one extreme and a thick and almost chorused sound at the other end. The width can be controlled by hand, or you can add a controller, such as a low-frequency oscillator (LFO) to alter the width over time (Figure 9.2).

The LFO is a special oscillator. It operates between one and ten hertz, which is too low to be heard by our ears. The LFO is not used to create waveforms but to alter them. LFOs are controllers because they can control oscillators and filters. We will look at controllers such as the LFO later in this section.

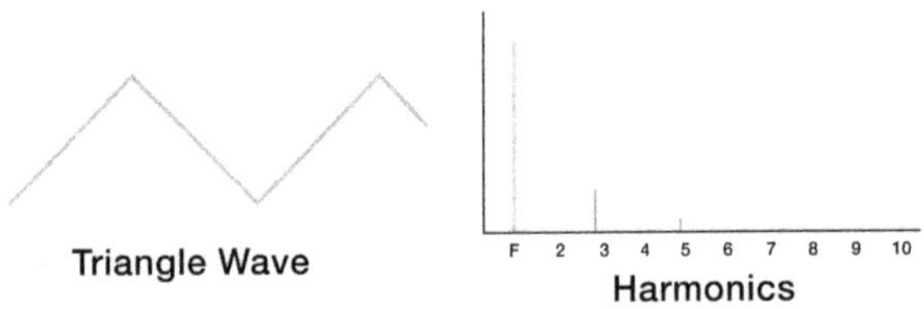

Figure 9.1 Three cycles of a triangle wave with harmonics

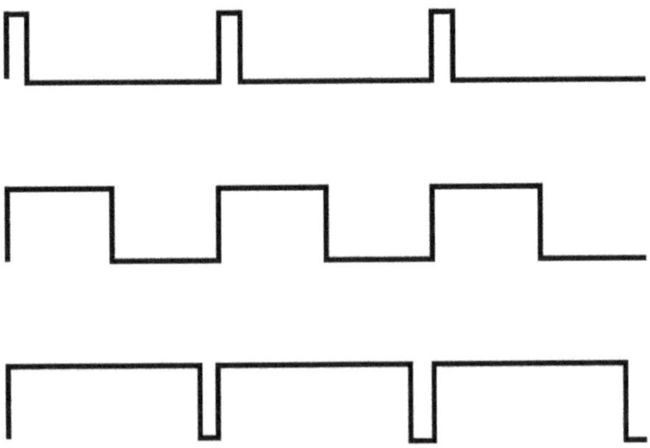

Figure 9.2 Three different shapes of a pulse wave

Noise

Another type of sound generator is the noise circuit. The noise circuit produces random electronic noise. Electronic noise is created by playing all frequencies at the same time and varying the loudness of each frequency randomly. Noise generators are useful in analog synthesis because they can add random frequencies that can be shaped with filters. Noise can be mixed with a waveform. The additional harmonics from the noise can be shaped with filters and change the overall timbre of the waveform. Noise generators do not alter the shape of the waveform, but the additional frequencies add color and variation to the sound.

Noise generators create white and pink noise. We looked at these two noise types in Chapter 2. White noise is made up of mixed, random frequencies with all frequencies being played at the same intensity. Pink noise is also made up of mixed and random frequencies, but the intensity of the octaves decreases as the frequencies get higher. This results in a less aggressive sound than white noise. White noise can create wind effects. Pink noise can create ocean sounds.

Oscillator Mixing

To increase the sonic potential of an analog synthesizer, manufacturers developed synthesizers with up to three waveform oscillators. The additional oscillators allowed you to combine different waveforms or stack the same waveform to create a thicker and bigger sound. A simple audio mixer was added to the synthesizer to control the balance between the multiple oscillators and the noise generators.

Stacking the same waveform two or three times created a bigger sound, but there was a way to create an even larger sound. Instead of having all three oscillators tuned identically, detuning the pitch of the second and third oscillators flat or sharp created a chorusing effect, which made the sound larger. This technique was popular when creating bass sounds on analog synthesizers (Figure 9.3).

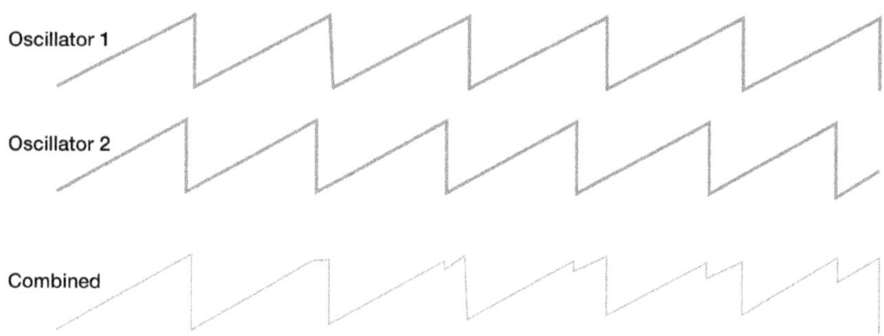

Figure 9.3 Result of two sawtooth waveforms slightly detuned

Controlling Pitch

To control the pitch of an oscillator, Moog attached an electronic keyboard to send specific voltages to the oscillator. Each volt would increase the pitch by an octave up. Moog used one volt as the lowest pitch, for example, A2. Two volts would create the pitch A3, three volts A4, and so on. The volt would be divided by 12 to create the pitches in between each octave. The pitch voltage output from the keyboard connects to the pitch voltage input of the oscillator.

The keyboard also sent out a separate voltage to open and close the amplifier called the gate output. Pressing down on any key of the keyboard would send voltage to open the amplifier so that we can hear the output of the oscillator. The amplifier would close when the key was released.

Since pitch was controlled by voltage, early analog synthesizers did not respond to Musical Instrument Digital Interface (MIDI) commands. When Dave Smith began working on the Universal Synthesizer Interface (USI) his first task was to develop a microchip that would convert voltage information into serial MIDI data. If one volt created the pitch A2, then the microchip needed to convert that voltage into the MIDI data to generate MIDI note number 45. Some manufacturers created kits to add MIDI to their analog synthesizers.

Sound Processors

Sound processors alter the waveforms and noise. They take the initial waveform(s) and alter the resulting sound. Sound processors consist of filters and amplifiers.

Filters

A filter is a circuit that can enhance or remove frequencies. A filter cannot enhance frequencies that do not exist. A filter can only affect frequencies that exist within the waveform. Analog synthesizers use several different types of filters.

An LPF allows low frequencies to pass while higher frequencies are removed depending on the cutoff frequency. A HPF allows high frequencies to pass while lower frequencies are removed depending on the cutoff. A bandpass filter combines the LPF and HPF. A bandpass filter only allows sounds on either side of the cutoff frequency to pass through. A bandpass filter allows you to remove low and high frequencies and only keep what is in the middle. A notch filter is the opposite of a bandpass filter in that it removes all the frequencies in between the two cutoff points (Figure 9.4).

Some filters feature a resonance control that enhances frequencies at the cutoff frequency. Resonance boosts the area around the cutoff frequency and emphasizes those frequencies. This slight boost adds additional harmonics to the overall sound and alters the timbre of the sound even further (Figure 9.5).

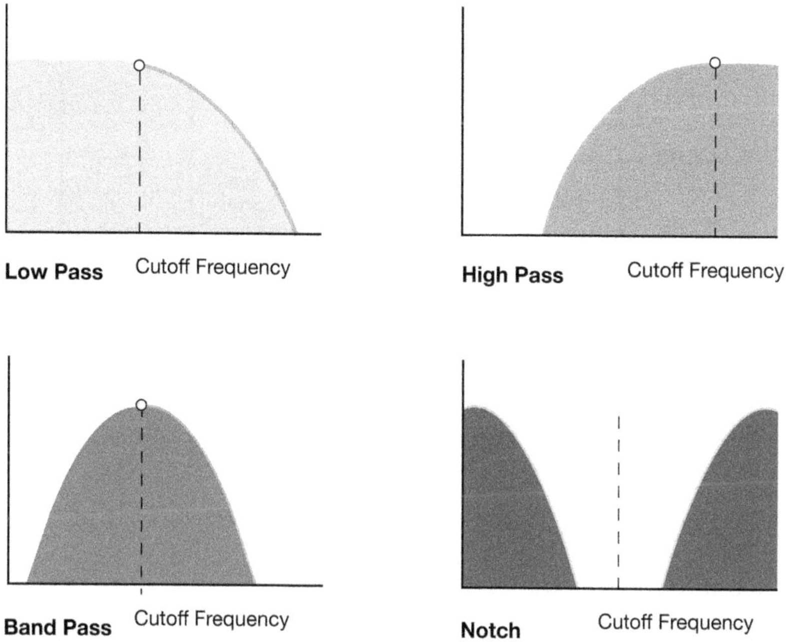

Figure 9.4 Four types of filters: low-pass, high-pass, bandpass, and notch

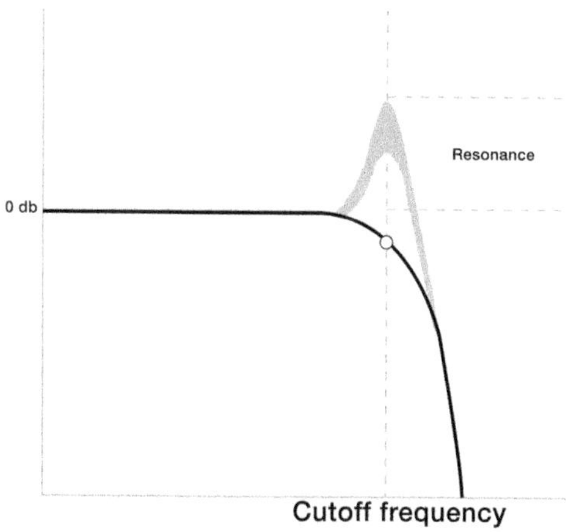

Figure 9.5 Resonance shown on a low-pass filter

The angle at which these filters begin to remove or enhance frequencies is determined by the filter slope. The filter slope removes frequencies gradually, which is less dramatic than a vertical line and sounds more natural to our hearing. The filter slope determines the steepness of the cutoff frequency.

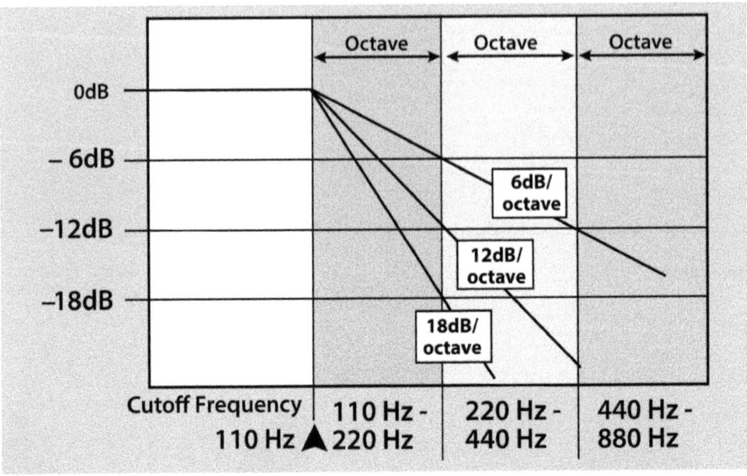

Figure 9.6 Common filter slopes with descriptions in dB/octave and pole

The angle of the filter slope is determined by how many decibels each octave is reduced. The greater the reduction, the steeper the angle of the cutoff. Common slopes are two-pole or four-pole, which refer to 12 dB per octave and 24 dB per octave. The image given in Figure 9.6 shows the different filter slopes found on analogy synthesizers.

The filters developed by Moog are considered by many as the best sounding in the industry. Moog's filters generate strong harmonics which create "rich" and "warm" tones. Moog did not patent any of the components or design of his synthesizer except for his filters. If you want the sound of a Moog filter, then you need to purchase a Moog synthesizer.

Amplifiers

An amplifier is a very simple circuit and all it does is increase the overall loudness of the waveform(s). An amplifier achieves this by increasing the voltage of the sound, thus increasing its overall loudness. Without the amplifier, you would have a very difficult time hearing the sound. Amplifiers are either on or off and are controlled by the gate output of the keyboard. If you want a sound to gradually fade in, then you need an EG to control the amplifier.

Controllers

A controller is a device that alters a sound generator or a sound processor. Controllers can be external sources or internal sources. Internal sources are integrated into the synthesizer such as an EG and an LFO. External controllers are music keyboards which are often equipped with dials and wheels.

Envelope Generator

We discussed a sound envelope in Chapter 2 and an EG performs the function on an electronic signal. The four parameters are identical: attack, decay, sustain, and release. The attack parameter determines the time it takes for a signal or sound to reach its full amplitude. The decay determines the time that it takes to signal to lower from its initial amplitude to its sustained amplitude. The sustain parameter determines the level of the sustained amplitude. The release parameter determines the time it takes for the sound to fade away after the sound has been released.

To control an amplifier with an EG, you connect the gate output of the keyboard to the gate in of the EG. Then, you connect the gate out of the EG and connect it to the gate in of the amplifier. The envelope begins the moment you press down on the key of a synthesizer and ends the moment you release the key. EGs are typically represented by the image given in Figure 9.7.

The EG primarily modifies the amplifier. Remember that an amplifier is either on or off. An envelope shapes our sound by controlling how the amplifier opens and closes. We can set our sound to have a fast attack and release, or a slow attack and quiet sustain level, or we can have a long release where the sound rings out for several seconds after we release the key. The EG has four dials or sliders to control the four parameters on a synthesizer.

The EG can also control the frequency or resonance of a filter. The EG can sweep the cutoff frequency up and down. A slow attack will raise the cutoff frequency slowly. We then set the decay and sustain to hold the frequency at one level until we release the key and with a long release the cutoff frequency will slowly return to its original frequency. We can do the same with the resonance setting to increase the amount of resonance and then return it to its original

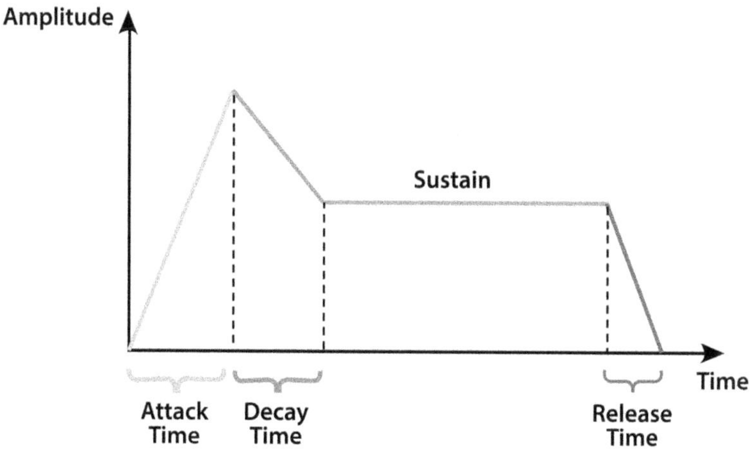

Figure 9.7 The four stages of an envelope generator

level. When using an EG to control the filter, there might be situations where we want to lower the cutoff frequency instead of increasing it. Many synthesizers have an inverse feature for the EG where it will operate in the negative direction. Instead of going up the attack goes down. This adjustment gives you greater control of the EG to modify the filter even further.

Low-Frequency Oscillator

The LFO is an oscillator that operates at an extremely low frequency, usually between 1 Hz and 10 Hz. LFOs are not meant to the heard: they are designed to offer you another of controlling parameters on a synthesizer. An LFO is typically a sine wave, but some manufacturers will allow you to choose from other waveforms such as a sawtooth or square. Since it is an oscillator, the LFO provides a repetitive signal for controlling sounds.

An LFO can be used to alter the pitch of an oscillator to create a vibrato or tremolo effect. When using a sine wave, the pitch of the oscillator will go up when the sine wave goes up and drop when the sine wave goes down. The frequency of the LFO controls how fast the change occurs. Most synthesizers give you a control on the oscillator to adjust how much the LFO affects the pitch. The LFO can also control the cutoff frequency or resonance of a filter.

If you want to add more variation to your sound, you could use an EG to control the frequency of the LFO over time. As the frequency of the LFO increases, the rate of pitch change increases on the oscillator. Using an EG allows you to have the LFO frequency change up or down quickly or slowly. This could make the vibrato effect on your oscillator less predictable.

External Controllers

The keyboard is the most common external controller since it allows you to control the pitch of the synthesizer and use it as a musical instrument. The keyboard also triggers the EG or amplifier to open and close. Keyboard controllers often have two common dials or wheels to further shape the sound. The first is pitch bend. The control enables you to temporarily bend the pitch up or down by several steps, usually a fifth or an octave. The second control is modulation. This control can be mapped to several parameters on the synthesizer. A common use is to control the frequency of an LFO controlling the pitch of an oscillator. Another one is to change the frequency of a filter. Just like other controllers, you can have the modulation wheel control several parameters.

The beauty of working with analog synthesis is that while it may seem limited by design, you have the freedom to mix and match the sound generators, sound processors, and controllers to create a variety of sounds that are not static and not boring. In the activity section of this chapter, we will look at a free synthesizer and test out the sounds that we can create with it.

Analog Synthesizers

Below is a list of classic analog synthesizers. These synthesizers are noteworthy because of their sound and popularity. These synthesizers are collectors' items and can be very expensive. Fortunately, all the synthesizers listed here can be found in virtual versions from several manufacturers. You can learn more about these synthesizers by visiting sites like www.vintagesynth.com. Each of the synthesizers listed here is available as a virtual instrument from different manufacturers.

Moog Modular Synthesizer

The original Moog synthesizer consisted of individual modules for each component mounted into a case. The number of oscillators, filters, and other components was limited only by your budget. The system had no memory to recreate sounds which meant writing down each setting and patch connection.

Moog Minimoog

The Minimoog is a portable, monophonic synthesizer. The modules were interconnected so patch cables were not required. The signal path was simple and consistent. It featured three oscillators and created big sounds and punch bass lines. This all-in-one until featured a 44-note keyboard.

ARP 2600

This monophonic synthesizer was semi-modular. You could create sounds without patching, but you had the option to patch if you wanted more flexibility or change the signal routing. The synthesizer was built into a portable case. A separate keyboard and speaker could be connected to the unit. The synthesizer was very popular for creating drum sounds and sweeping pads.

Oberheim SEM

The Synthesizer Expander Module (SEM) was a single module with two VCOs, two filters, and two EGs. Oberheim created their own protocol that allowed you to connect multiple SEM units to create a larger system. Adding modules meant you could create a polyphonic synthesizer controlled from a single keyboard.

Oberheim Matrix-12

This was Oberheim largest and biggest sounding synthesizer with two oscillators per voice and 12-note polyphony. The unit featured multiple filters and 15 LFOs. The synthesizer featured a modulation matrix that allowed you to modify nearly any parameter from any number of controllers. This meant you could create expressive sounds that responded to several parameters.

Yamaha CS-80

This 200-pound analog synthesizer is considered Japan's first great synthesizer. The synthesizer featured eight-note polyphony and a five-octave weighted keyboard with velocity and aftertouch. The oscillators created large brass and string sounds. This synthesizer had patch memory to store and recall programmed sounds.

Sequential Circuits Prophet-5

The Prophet-5 is one of Sequential Circuits' most popular synthesizers. The synthesizer featured two oscillators per voice with five-note polyphony. The filter had a different character compared to the others at the time, which led to unique sounds. The patch memory could memorize the exact location of the knobs, which made recalling sounds easy. This was one of the first synthesizers to offer a kit to add MIDI.

Roland Jupiter-8

Another popular Japanese synthesizer featured in countless recordings in the 1980s. This synthesizer was Roland's first professional synthesizer. The unit offered two oscillators per voice and eight-note polyphony. The synthesizer could store all patch information and had a five-octave keyboard. A separate LPF and HPF allowed greater sound programming options. Later models offered a MIDI retrofit kit.

Roland Juno-106

Roland's first synthesizer to use a digitally controlled oscillator (DCO) instead of a VCO. The digital oscillator meant that the waveforms and pitch were controlled digitally. This allowed the synthesizer to include MIDI as standard. The DCO was stable but lacked the depth and character of a VCO. With only one DCO, it was not capable of creating large sounds like the Jupiter series. It used the same filters as the Jupiter synthesizers, so it had the potential to create interesting sounds. The synthesizer had a five-octave keyboard but offered no velocity.

Digital Synthesis

Digital synthesis describes any synthesis technique that does not use analog waveforms as its sound source. Digital synthesizers can create new and unusual waveforms and lead to a greater sonic variety. Digital synthesis creates its waveforms using algorithms and computer processing. Waveforms are created each time a key is pressed or can be pulled from a collection of waveforms stored on the synthesizer.

After creating the waveform, digital synthesis follows a similar signal path as analog synthesis. The waveform can be modified by a filter, and then pass through an amplifier modified by an EG. A digital synthesizer can also have an LFO to modulate the waveform pitch and filter. Digital synthesis can offer more control options depending on how the controllers and destinations are programmed. There are a lot of possibilities with digital synthesis, and we will explore some of the more popular and significant methods.

Frequency Modulation Synthesis

For many, the first digital synthesizer was the Yamaha DX-7. It was not the first digital synthesizer, but it was the first commercially successful digital synthesizer. The DX-7 had 16-note polyphony, greater than any analog synthesizer. It could produce bell and brass tones that analog synthesizers could not. The 61-note keyboard had velocity and aftertouch. MIDI implementation was standard, something that few analog synthesizers offered. It was difficult to program, but that did not matter because it produced sounds not heard on an analog synthesizer.

Frequency modulation (FM) synthesis involves at least one waveform modulating the frequency of another. The first oscillator is called the carrier because it is responsible for the pitch that is produced. The second oscillator is called the modulator because it modulates the frequency of the carrier. Analog synthesis uses a form of FM when an LFO is used to modulate the pitch of an oscillator to create a vibrato or tremolo effect. On the DX-7, the frequency of the second waveform modulating the first waveform is significantly higher (Figure 9.8).

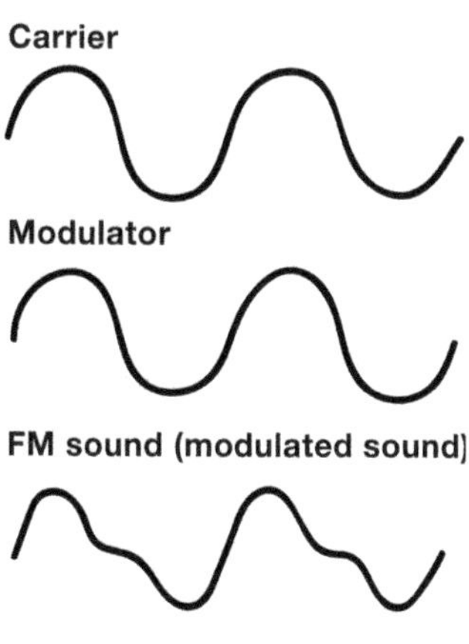

Figure 9.8 Carrier waveform modulated by the modulator waveform and the resulting waveform

John Chowning

The FM synthesis method was developed by John Chowning in the 1970s at Stanford University. Chowning was looking for a way to create new waveforms that were harmonically for synthesis since analog synthesis had limited options. He used computers to generate digital sine waves for his experiments and soon developed an algorithm for FM synthesis. His efforts led to unique sounds such that Yamaha took an interest in the technology. Chowning licensed the algorithm to Yamaha in the mid-1970s. Yamaha used the algorithm to develop a new digital synthesizer for retail sale.

Yamaha DX-7

Yamaha spent several years adapting Chowning's algorithm. Chowning's algorithm allowed for an infinite number of waveform combinations. Yamaha settled on 32 waveform combinations. Chowning's algorithm could use any number of operators, or waveform generators, and the possibility of using waveforms other than sine. Yamaha allowed up to six operators, or waveform generators, and limited all the waveforms to sine waves. To add flexibility and control, each operator had its own EG and amplifier.

The DX-7 was an enormous success when it was released in 1983 and remains one of the best-selling synthesizers of all time. Even though Yamaha simplified Chowning's algorithm, programming the synthesizer was extremely difficult from the front panel. The panel consisted of a small LCD screen and numerous buttons to navigate menus. Several software manufacturers wrote editing software so that you could program the synthesizer from your computer. Even though the interface was difficult to use and program, users purchased the synthesizer because it produced bright tones such as bells and electric pianos which analog synthesizers could not. The DX-7 is available as a virtual instrument by several manufacturers.

Additive Synthesis

Additive synthesis began as a mathematical theorem developed in the late 18th century by a mathematician of the name Jean Baptiste Joseph Fourier (1768–1830). The theory states that any waveform could be defined as a collection of sine waves at different frequencies. When sine waves of different frequencies are combined, new harmonics are created which alter the timbre of the waveforms. This means that regardless of how complex your original waveform is, it can be defined as a series of sine waves at different frequencies (Figure 9.9).

This theorem led to the development of the Fast Fourier Transform algorithm. This algorithm analyzes a waveform and calculates the number of sine waves with specific frequencies and amplitude levels needed to create

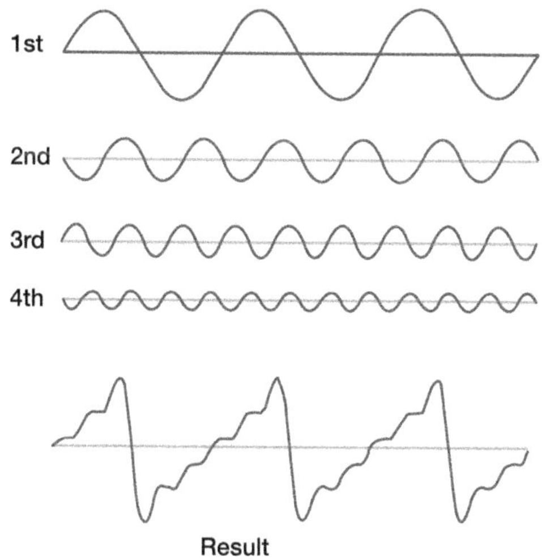

Figure 9.9 Four sine waves at different frequencies adding up to a sawtooth waveform

the waveform. Once the waveforms were calculated, a computer could be programmed to assemble the waveforms to recreate the sound. An additive synthesizer could theoretically recreate any existing sound, but the process required an immense amount of computer processing. Additive synthesis was seen as a way of creating new and complex waveforms not possible with analogy synthesizers.

The most accessible additive synthesizer in the market was the Kawai K5, which was released in 1987. The K5 generated up to 126 sine waves for sound generation. Programming was difficult because of the small screen and navigating through submenus. The synthesizer produced a variety of interesting and useful sounds that were not possible with analog synthesizers. This synthesizer is not available as a virtual instrument.

Wavetable Synthesis

Wavetable synthesis involves collection of single-cycle waveforms that can move from one waveform to another. You could start with a sine wave, and then transform it into a sawtooth over a period of time. As the sine wave evolves into a sawtooth, the timbre of the sound changes as the number of harmonics increases. Wolfgang Palm developed wavetable synthesis in the late 1970s. He refined the technique and eventually released a synthesizer called the PPG Wave (Figure 9.10).

The PPG Wave could cycle through a selection of waveforms so that the sound could evolve over time. The ability to change from one waveform to

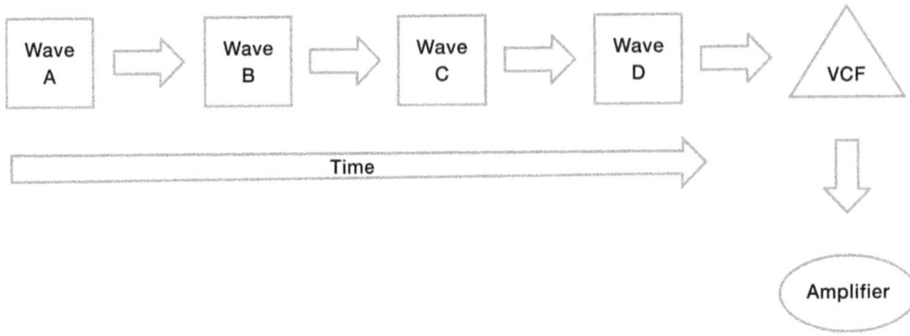

Figure 9.10 Signal path of a wavetable synthesizer

another was determined by variables such as time, velocity, or a modulator such as in LFO. The goal was to create sound that was not static and constantly evolved. Each waveform had its own EG and amplifier. The waveforms were combined through a mixer and then passed through a filter and amplifier to produce a final sound. Like analog synthesizers, the sounds could be modified with EGs, filters, and LFOs. The PPG Wave featured 30 waveforms and eight-note polyphony. The synthesizer featured a small screen which made programming very difficult. Although the synthesizer could produce very interesting sounds, the steep price meant it did not sell in large quantities. The PPG Wave is available as a virtual synthesizer.

Vector Synthesis

Vector synthesis is a variation of wavetable synthesis. Vector synthesis was developed by Dave Smith of Sequential Circuits in 1986 with the release of the Prophet VS. Smith began with a collection of 128 waveforms. The synthesizer could play up to four waveforms from the wavetable. A joystick on the synthesizer allowed you to move freely between the four waveforms so that you play individual waveforms or any combination. The path of the joystick could be altered by modulators such as velocity, pressure, two LFOs, and the modulation wheel. The result was a synthesizer capable of producing unique and very interesting sounds. Programming was difficult due to a small display screen and like the PPG Wave, the pricing was not competitive, and the Prophet VS did not sell well. This synthesizer is available as a virtual synthesizer.

Granular Synthesis

Granular synthesis describes any digital synthesis technique that uses "grains" from a sound as a starting point. A grain is an extremely small particle of sound that lasts between 10 and 100 milliseconds in length. Instead of using a single cycle from a waveform, you would use a random collection of grains

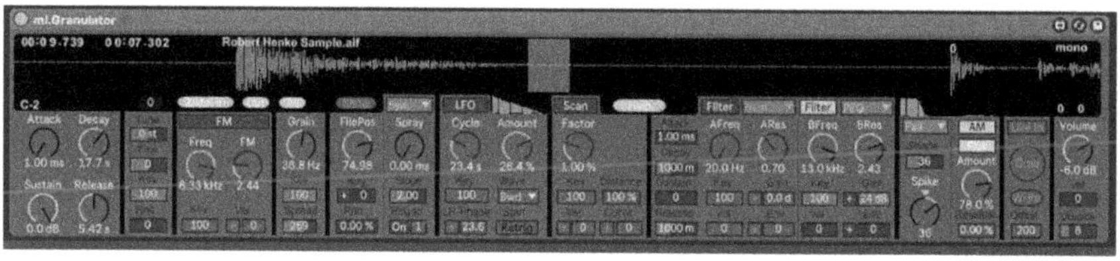

Figure 9.11 Granulator II interface in Ableton Live (Courtesy of Ableton AG, www.ableton.com)

from the sound. The grains are usually large enough to capture the pitch and the harmonic content of sound, but not long enough to identify what the source of the. This is perhaps the most compelling part of granular synthesis; you can take existing sounds and turn them into completely different and unrelated sounds.

Once the grains are collected, you can control how many of the grains will sound, the order they sound, and how fast each grain is played. These variations provide an incredible range of sonic possibilities. The implementation of granular synthesis differs from each manufacturer, however, the basic principle using grains is common between them. The resulting sound from granular synthesis is unpredictable but the technique is useful in creating evolving textures. Because of the computing power required for granular synthesis, most of the synthesizers are software based. One example is Granulator II, which is available in Ableton Live (Figure 9.11).

Virtual Instruments

I stated that the analog and digital synthesizers we explored are available as virtual instruments. Now, we need to address what a virtual instrument is and how they are created. A virtual instrument is a computer-generated version of a real instrument. The virtual instrument should look, sound, the function the same as a real instrument. Virtual instruments are convenient because they run inside our computers and do not occupy physical space. Virtual instruments can be created so that they exceed the limitations of the actual instrument. For example, the Minimoog is a monophonic instrument. The Minimoog virtual instrument has unlimited polyphony. Virtual instruments open a world of possibilities.

Physical Modeling

The goal of physical modeling is not to capture the sound of an instrument but rather recreate the sound exactly as it is created on a physical instrument. To create a physical model of a snare drum, one must study how a snare drum creates sound. This means you must think about how the stick strikes the head and understand that the snare sounds different depending on whether you strike the edge or center of the head. You must calculate the vibrations in the material of the snare drum, for example, wood sounds different than that metal. You must even calculate how the sound changes depending on the force the drumstick strikes the head. A modeling synthesizer creates a new waveform each time the drumhead is struck.

The act of creating a new waveform each time a note is played requires a significant amount of computing power. The amount of computing power depends on the accuracy of the model. The appeal of physical modeling is that you have a system that responds to player input. Physical modeling ideally recreates the sound of the instrument without sounding repetitive or stale. Rather than playing a single sample of a snare drum, each strike is different from the previous one, adding to the realism of the physical model. The waveform of the sound is computed using mathematical equations. The programmer must create equations and algorithms that calculate the physical source of the sound. This means that the software must respond to how the player plays the instrument.

Virtual Digital Synthesizer

Creating models of acoustic instruments is complicated because of all the variables of acoustics involved. Synthesizers might seem easier, but they have their own complications as well. Early digital synthesizers were created on printed circuit boards. The waveform calculations for the Yamaha DX-7 were handled by a microprocessor on a circuit board and the sine waves were created digitally. The EGs and amplifiers function in the digital domain. The entire signal path was digital until the very end, when a digital-to-analog converter transformed the digital data into electrical current.

Electronics today are still delivered on printed circuit boards, but the design of the circuits is done on computer software. It stands to reason that one could take a digital synthesizer like a DX-7 and create a virtual version of the instrument that sounds the same as the original. However, there are small imperfections on the circuit boards that cannot always be replicated. One solution is to add a randomizer that will simulate imperfections in the circuits. The result is that the virtual instrument is not exactly like a DX-7, but it is close enough. If you isolate the electric piano on the DX-7 and compare it with the virtual one, you might find differences in the sound. However, once the instrument is playing in context with other instruments, the differences cannot be heard.

Virtual Analog Synthesizers

Analog synthesizers are more difficult to model because analog electronics are unstable and constantly vary. Digital synthesizers have predictable behavior because they are essentially computer programs; analog synthesizers are not. It is not as complex as an acoustic instrument, but there still are many variables. With analog synthesizers, programmers must model the behavior of the electronics. The behavior of a filter may be different with a sawtooth wave than it is with a square. This means the model must calculate how the filter behaves depending on the waveform.

Modeling a Roland Jupiter-8 synthesizer means studying the filter in multiple situations and then deciding how responsive the model is to waveform changes. A VCO playing a sawtooth waveform may not be consistent in harmonic content the higher the pitch; this is another behavior that must be studied. A model of an analog synthesizer may differ more from the original than a digital synthesizer. But considering the cost to purchase and maintain a Jupiter-8, the model is an excellent substitute.

Virtual Drums

Prior to virtual drums, if you wanted to add drums to your project without recording a drummer, you had only a couple of choices. You could find a collection of audio or MIDI drum loops and add them to a track in your project. The loops automatically conform to the project tempo. You do not have a lot of control with loops, so if you are looking for a specific pattern, you would need to preview each loop until you find one that worked. If you are comfortable with working with MIDI, you could create your own drum parts and then use a drum library to play back the sounds. However, not all of us know how to write drum parts.

At the core, virtual drums are a collection of drum samples and/or models that sound like acoustic drums. The interface might even show a drum set where you can load an entire kit or customize each drum piece. A virtual drum instrument can also include a collection of MIDI loops organized by style and phrase. This allows you to search for drum fills for a ballad, for example. This type of instrument is very useful provided the MIDI loop library has enough variety.

Some virtual drums offer variations or randomizers for patterns. You select a pattern by pressing a pad on the software, and then you can choose variations on that pattern and even control how many measures pass before a drum fill occurs. The variations on a pattern are subtle, but they do make a difference. You can start with an eight-bar pattern for the first verse of a song and have a drum fill at the end of the eight bars. You can then use a variation of that pattern for the second verse and have a different drum fill at the end. Then, you find a pattern for the chorus and add variations from there. This solution takes a little time, but you can create convincing results.

A more sophisticated solution that some DAW applications offer, begins with you labeling each section of your song such as the intro, verse, chorus, and outro with marker. You then load the drum instrument and choose a style for your song; we can stick with a ballad. The virtual drummer analyzes the song based on the markers and automatically adds patterns for each section. This is an extremely fast way to work, and if you do not like the results, you can try a different style.

The generated drum parts for these solutions may not be exactly what you want. However, if you can convert the parts into MIDI, you can always edit the drum parts to your liking. You can add additional fills or crash cymbals to signal important parts of the song. What is important is that you do not have a static drum pattern that does not change for the entire song. Drummers add slight variations to their patterns to keep the song moving forward. These solutions cannot necessarily replace a live drummer, but they do offer your options when recording a live drummer is not practical.

Virtual Enhancements

When modeling vintage synthesizers, programmers have the freedom to expand the original specifications. Some synthesizers did not have the ability to store patches. Others were limited to 64 or 128 patches. Modeled synthesizers can have unlimited patches. Monophonic synthesizers or those with limited polyphony can be expanded to have unlimited polyphony. Digital effects such as reverb, chorus, and delay can be applied to the signal path to add dimension to the synthesizers. Synthesizers that did not have MIDI implementation can now support MIDI. Finally, any noise or distortion present in the analog outputs can be removed, providing a cleaner sound. There are users who feel that the noise and imperfections contribute to the character of these instruments, so some manufacturers give you the option to enable and disable functions. Virtual synthesizers allow us to access vintage synthesizers that no longer exist or are extremely rare.

Sampling

A sampler is not a synthesizer because it does not create sound using synthesis techniques. A sampler attempts to re-create actual sounds by playing back a recording of that sound. Sampling is still a viable solution for playing back acoustic instruments on your computer. If you want to have the actual sound of a cello playing middle C, then record a cello playing middle C and then play back the recording when you need the sound. With digital editing, you could pitch shift the recorded note up and down several times until you have enough notes to play a scale. Pitch shifting the notes means that you do not have to make multiple recordings of the cello to play a scale.

The principle seems simple enough, however, there are variables to consider. For example, if you ask a cellist to play middle C four times in a row, it is unlikely that each repetition is identical. There are variations in bow pressure each time a note is played that affects the timbre. The attack of the note differs when it is played with an up bow as opposed to a down bow. If we play our sampled middle C four times, it will sound exactly the same each time, which our ears will immediately detect. To sample an instrument convincingly, we need to do more than just make a recording of a single note. Let us look at the steps of creating a sample library.

Sound Recording

The first stage of sampling is capturing or recording the sound. Let us continue with our cello example. The capture the sound, we need to place microphones in different places of the instrument. Our ears hear the instrument in a room, so we need to capture the sound in the room as well. If we have the storage capacity to record each note on the instrument, then we can multisample the instrument. If we need to save space, we capture every fourth or fifth note and then pitch shift or transpose the samples up and down to fill in the missing notes. Each transposed sample is mapped to a few notes on either side of the original sample, creating a keyzone. We must be careful when transposing samples up and down because if we go too far, it will sound artificial (Figure 9.12).

Articulations

A cello has different articulations, that is, different ways for the note to start sounding. Our cello library can start with a legato articulation with each note

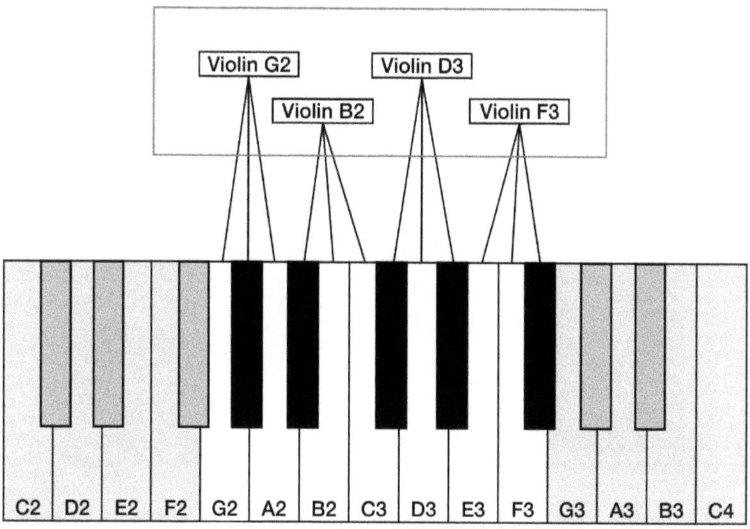

Figure 9.12 Notes that make up a keyzone

lasting three seconds. The person playing the cello has to be precise so that each legato note played has the same attack and sustain. If we want another articulation, such as a pizzicato, which is when the note is plucked with a finger, then we need to make more recordings. Once again, we can multi-sample the cello with a unique sample for each note, or we can record fewer notes and create keyzones by transposing the samples up and down.

Switching articulations from legato to pizzicato meant loading all the legato samples and then loading the pizzicato samples. To make the switch easier, keyswitching was developed. Keyswitching assigned an extremely low or high note to switch between articulations. Pressing the note would load the other articulation until you pressed another key to switch back to legato. Keyswitching required more memory since more samples need to be loaded, but it provided a convenient way to switch between different articulations (Figure 9.13).

Sustained Notes

Our legato articulation has each note lasting three seconds. After three seconds, the note stops playing, even if we continue to press the key down on the keyboard. We could record a legato articulation where the note lasts 15 or 30 seconds, but that requires even more disk space. Another option is to loop the three-second sample so that it continues to sustain until we release the key. Creating a loop involves opening each sample in an audio editor and looking for a portion of the wave that will loop smoothly. A smooth loop will have no audible pops or clicks and will not suddenly change the volume.

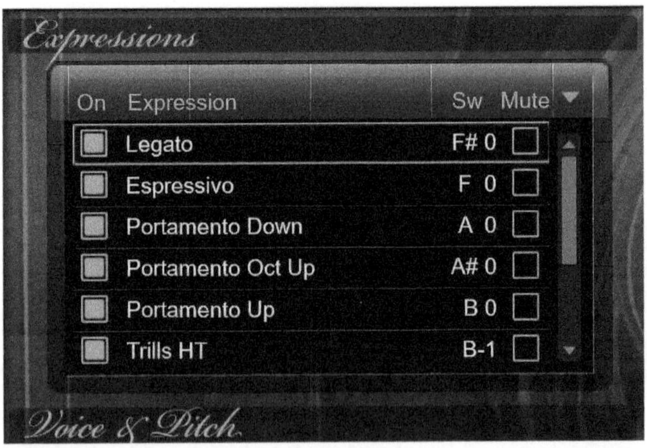

Figure 9.13 Notes assigned on a keyboard for keyswitching in Halion Symphonic Orchestra (Courtesy of Steinberg Media Technologies GmbH, a division of Yamaha Corporation, www.steinberg.net)

The best place to find the loop is after the decay and before the release of the sound. Looping saves storage space, but if not done carefully, can take away from the realism of the instrument (Figure 9.14).

Dynamics

A cello can play loudly or softly. The note will sound different depending on the force the player uses. A loud note tends to have more harmonics while a quieter one has less. When multi-sampling, we record the same note at different volumes. We then assign velocity levels to each volume. If we take five different samples at different volumes, the first volume could have a velocity range of 1–25, the second 26–50, the third 51–75, the fourth 76–100, and the last 101–127. Assigning a different sample for different velocities is called velocity switching. If I press the key hard, then the last sample plays. If I strike it very gently, then the first or second sample plays. Keep in mind that this process takes up more storage space and if we are looping our samples, then each sample layer must be looped (Figure 9.15).

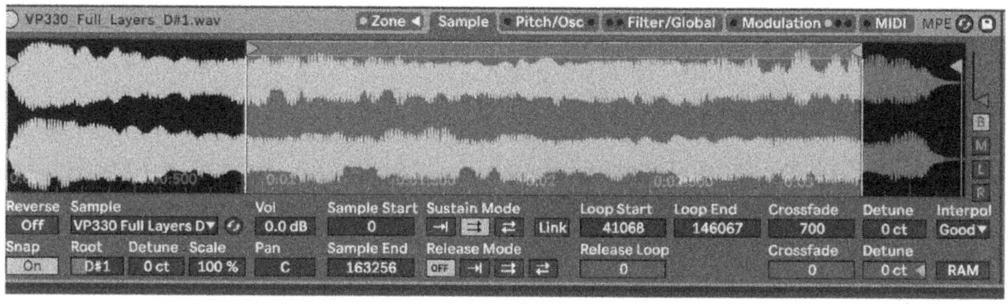

Figure 9.14 Looped waveform in Sampler by Ableton (Courtesy of Ableton AG, www.ableton.com)

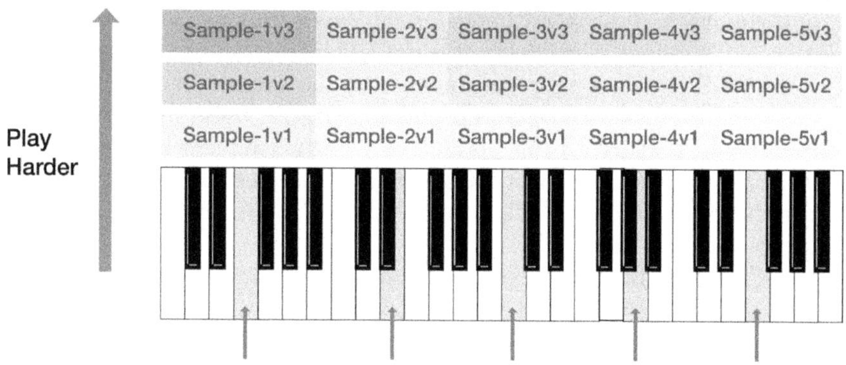

Figure 9.15 Velocity switching on a keyboard

If we need to save storage space, then we can try to simulate quiet notes. We record the cello playing legato at a strong dynamic level, say 100 in velocity. We could play back that sample quieter when pressing the keys lightly. However, it will be obvious that we are listening to a loud sample at a quieter level. One solution is to place an LPF on the sample and have the cutoff frequency respond to velocity. When the note is played strongly, the cutoff frequency is higher. At lower velocities, the cutoff frequency is lower. The technique can help simulate the change in harmonics between a loud note and a soft one.

Sample Library

Once we have all our samples, we need to build our sample library. The sample library contains all the samples needed for our instrument. The samples are loaded into a sampler, which then manages all the sounds. To save space, some samplers use lossless compression to reduce the file size. Other samplers will lower the sample rate and bit-depth to create smaller files. This reduction could compromise the audio fidelity. Once the files are packaged, then the sampler can play the samples in the library. Samplers can be a MIDI hardware devices, a software application, or a plug-in.

Hardware Samplers

In 1981, E-Mu Systems revealed the first commercial digital sampler called the Emulator I. This sampler had eight-note polyphony with a sampling rate of 27.7 kHz and 128 kilobytes (kb) of memory. This amount of memory seems extremely small today, but at the time it was significant. The Emulator II was released in 1984 and increased the sample rate to 32 kHz and memory to 512 kb. Later models would add a CD-ROM drive and an internal hard drive. The Emulator II was an expensive sampler at the time, which meant it did not sell in great numbers.

In many ways, these were simple instruments with keyboards and limited memory and storage. The fidelity was not CD quality and the sample libraries were small and not very convincing. All the samples had to load into memory before you could play the instrument. Yet these instruments had character and that led to their popularity. Early samplers had samples of acoustic instruments, but most users were more interested in recording their own samples and creating unique sounds, such as a percussion instrument made up of brake drums.

The Emulator II sampler offered analog filters and multiple LFOs for extensive sound shaping. The string sounds, for example, did not sound like real instruments, but they had their own character and sound, which made them usable in a variety of scenarios. The popularity of the Emulator II was the ability to edit samples and create unique sounds. The low fidelity of 32 kHz and eight-bit depth added to the personality of the instrument. E-Mu also released a drum machine sampler called the SP-1200 with similar specifications to the Emulator II. The SP-1200 is still sought after and considered one of the best sampling drum machines ever released.

Companies like Akai and Roland would later offer samplers with more features and storage at a lower cost. Their sounds and character could not compare to the offerings from E-Mu Systems. Fortunately, the Emulator II sample library is still available for software samplers and there is a virtual instrument version of the Emulator II that features most of the original sample library.

Software Samplers

By the mid-1990s, digital sampling was available on computers, however, all samples still needed to load into memory, which meant large libraries were not possible. In 1996, 32 MB of RAM was considered extravagant because of the cost. In 1998, a software developer named Nemesys released a software sampler called Gigasampler. Gigasampler was written so that samples could be streamed directly from the hard drive and did not need to be loaded into the computer memory. This ability completely changed the sampling industry because now it was now possible to have extremely long samples and more dynamics. Instead of having a looped cello, you could play a long sample of a cello holding a note for 30 seconds or longer. Using hard drive storage meant more articulations using velocity and keyswitching. There were limitations to Gigasampler, however. The software was standalone, which required users to have two computers, one for their DAW and the second for Gigasampler. The Gigasampler computer required a specialized audio interface and a MIDI interface to receive MIDI from the DAW. Finally, the software only ran on Windows computers. For many users, the benefits of Gigasampler were worth these limitations.

By the end of the 1990s, other companies developed sampler plug-ins that offered hard disk streaming and could read Gigasampler libraries. For a variety of reasons, Nemesys could not deliver a plug-in version of their software and eventually, the company went out of business. Today there are several manufacturers offering extensive libraries featuring full orchestras and choirs. Composers can now create extremely convincing mock scores of their music with ease. Songwriters can add strings or an acoustic piano to their songs without needing to go to a studio. Some choir libraries let you type in lyrics and the samples will sing your lyrics. There are other libraries that feature ethnic instruments, which are useful for composers needing to add authenticity to a documentary.

Romplers

Romplers are sample-playback synthesizers. The term rompler is a combination of ROM, which stands for read-only memory, and sampler. As the name implies, this synthesizer can play back samples, but could not record sound or create samples. Romplers were seen as a solution for those who wanted the ability to work with samples but did not want the expense or complexity of a dedicated sampler. Romplers were often multi-timbral, which offered musicians greater flexibility when creating multiple parts. Romplers were often rackmount or desktop units without keyboards, which helped lower their cost.

Several companies created romplers, most notably E-Mu with the Proteus series. The Proteus romplers came in different formats with a variety of samples. There were units devoted to orchestral sounds, world sounds, percussion, modern sounds, and even synthesizers. One could build a virtual orchestra by combining multiple Proteus units. The samples were high quality, and the interface was easy to use. Roland, Yamaha, and other companies followed with their own romplers. Romplers were extremely popular in the 1990s and remained so until virtual instruments appeared in DAWs and offered the same quality and variety of sounds without requiring hardware devices.

Summary

In this chapter, we spent time looking at different types of synthesis techniques: analog, FM, additive, wavetable, vector, and granular. We learned that each of these techniques produce different results and sounds. We discovered that these different techniques produce waveforms differently, but that they all follow a similar signal path. We learned about modeling synthesis, and how this technique allows us to have virtual versions of vintage synthesizers in our sessions. We closed with samplers and how they help us add acoustic instruments such as pianos and orchestras to our music without the expense of hiring musicians and renting a studio. We have just scratched the surface with these methods and there are plenty of instruments and sample libraries to explore. There are a lot of sound options available to you, but start with what you have in your DAW, and grow from there.

Activity 9

Analog Synthesis

In this activity, we will learn how to program a virtual modular synthesizer. Cherry Audio offers a selection of competitively priced virtual synthesizers. They also offer a basic virtual modular synthesizer called

Voltage Modular Nucleus for free. This synthesizer will run on most Windows and Mac computers and does not require an audio interface. It will work with any sound card on your computer.

Install Voltage Modular Nucleus

Visit https://store.cherryaudio.com/bundles/voltage-modular-nucleus to download the Voltage Modular Nucleus virtual synthesizer. You will need to create an account with Cherry Audio to "purchase" the product for free. After you create your account and purchase the synthesizer, you can download the appropriate version for your computer. Voltage Modular Nucleus will run as a standalone application, as well as a plug-in. If you do not have a DAW application on your computer, you can still run the synthesizer and complete this activity. The instructions assume you are running the synthesizer in standalone mode.

Verify Settings

When you first run Voltage Modular Nucleus, the application may update itself and install the libraries. Once it is complete, you will see the empty interface on your screen. Click on the gear icon toward the top center of the screen and then click on the Audio/MIDI button. The synthesizer automatically selects the default sound card for your computer. Click the X in the top right corner to close this window.

We can verify that we have sound by going to Presets at the top left and selecting a preset. To keep things simple, click on the Basic category and select Basic Sawtooth. The window will show various modules. To play the synthesizer, click on the keyboard icon to open the virtual keyboard. The virtual keyboard allows you to use your computer keyboard to play notes. Before you do so, you may want to turn down your speakers or headphones, sometimes the output from this synthesizer is loud. Once you verify that you have sound, click on the New button at the top left to load a blank page. Now, we will begin to program our own sound.

Connecting the Amplifier

The left column has a series of modules to choose from. We are going to start with the Amplifier module. Locate the Amplifier module and drag it to the workspace in the center of the screen. You will need to connect the positive output on the amplifier to the 1L (M) connection near the top right under MAIN OUTS. To connect the two modules, click and hold the positive output port on the amplifier and then start dragging up while still holding the mouse down. A cable will appear. Connect the cable to the 1L (M) connection. Any sound that comes into the amplifier will be heard on our computer speakers. Under MAIN OUTS you may want to turn the volume down so that it is not too loud when you start to play the synthesizer. I set mine to -10 dB.

We need to tell the Amplifier to open whenever a key is pressed. Go to the top left under CV Sources. Connect the Gate under CV Sources to the CV Amount on the Amplifier. Click and drag as you did when connecting the amplifier to the main outputs.

Adding an Oscillator

Now, we will add a sound source. In the left column look for the Oscillator module and drag it to the left of the Amplifier in the workspace. Under CV Sources connect the Pitch to the Pitch CV connector on the Oscillator. The Oscillator has several waveforms for us to choose from. We are going to use the sawtooth waveform which is the second one from the left. Connect the output sawtooth waveform to the input of the Amplifier. Now, when you press the keys on your computer keyboard you should hear a sawtooth wave.

Adding a Filter

Let us add a filter to help shape the sound of the oscillator. In the left column, locate the Filter and drag it to the right of the Oscillator. We now need to route the output of the Oscillator to the input of the Filter. Grab the wire that is connected to the input of the Amplifier and drag it to the Audio In connector of the Filter. Let us see how a high pass filter affects our sawtooth waveform. The high-pass filter is located at the bottom of the filter and is the first one from the left. Connect the output of the filter to the input of the Amplifier. Press a key on your computer keyboard to verify that you are still getting sound.

Drag the Cutoff knob on the filter until in the ten o'clock position. Press a key on your computer keyboard and listen to how the tone of the sawtooth waveform has changed. Continue to hold the note while you rotate the Resonance knob. Notice how the tone of the waveform changes even more and that you start to hear additional harmonics the more you turn the Resonance to the right. Feel free to experiment with adjusting the Cutoff and Resonance to alter the tone.

Adding an Envelope Generator

Now, we are going to add an EG to help control the amplifier. In the left column, look for EG and drag it to the left of the Amplifier. Grab the cable connecting to the CV Input of the amplifier and connect it to the Gate In of the EG. At the bottom of the EG, connect the output of the positive envelope shape, which is located at the right and connect it to the CV Input of the Amplifier. Now, adjust the Attack of the EG to about 450 milliseconds. Set the Decay to 450 milliseconds as well. Set the decay to about 1,000 milliseconds. Press the key on your computer keyboard and notice that the waveform takes a moment to fade in and continues to sound momentarily after you release the key.

Conclusion

There are several tutorials on how to make additional sound with Voltage Modular Nucleus on the Cherry Audio website. You should also load some of the presets to see the different types of sounds you can create with this virtual synthesizer. I hope this activity has helped you understand the simplicity and flexibility of analog synthesis.

This concludes this activity. If you want, you can save the project or quit without saving.

Terms

Additive Synthesis Synthesis method that adds sine waves to create waveforms with unique harmonic content.

Analog Synthesis Synthesis method using simple waveforms that are altered harmonically with filters. Also called subtractive synthesis.

Bandpass Filter A filter that only allows sounds on either side of the cutoff frequency to pass through.

Carrier Waveform The waveform that carries the pitch in FM synthesis.

Fast Fourier Transform A computer algorithm that analyzes a waveform and calculates the number of sine waves with specific frequencies and amplitude levels needed to create the waveform.

Frequency Modulation Synthesis Synthesis method that uses one waveform to modulate a second waveform to create new waveform shapes.

Grain An extremely small particle of sound that lasts between 10 and 100 milliseconds in length.

Granular Synthesis A synthesis method where grains of sound are combined and processed to create evolving textures of sound.

Keyswitching A sampling playback method where certain notes on the keyboard will load different articulations for an instrument.

Keyzone A collection of samples created from a single sample.

Low-Frequency Oscillator An oscillator that operates between one and ten hertz. Used as a controller in analog synthesis.

Modulator A waveform that modulates the pitch of another waveform in FM synthesis.

Notch Filter A filter that removes all the frequencies in between two cutoff points.

Oscillator An electronic circuit that produces a repetitive electronic waveform.

Physical Modeling Synthesis A synthesis method that attempts to recreate acoustic waveforms based on calculations derived from a physical instrument.

Pulse Waveform A waveform with a variable width ranges from a tall rectangle to a square and then to a wide rectangle.

Resonance A feature on a filter that boosts the area around the cutoff frequency and emphasizes those frequencies around the cutoff.

Rompler A sample playback synthesizer loaded with sounds into memory. A rompler cannot sample new sounds.

Triangle Waveform A waveform shaped like a triangle with similar harmonic content as a square waveform.

Chapter 10

Computer Music Notation

Music Notation

Computer music notation is the contemporary way of creating professional grade scores and parts for musicians. There are several software options that enable you to create professional music scores with ease. To be effective and useful, the software must be able to accommodate a variety of score types and layouts. The software must be able to import different file types to create scores. The software must also allow for different note entry methods to accommodate different users. The software must also deliver music scores in different formats aside from printing on paper.

Many digital audio workstation (DAW) applications offer integrated notation features, which may be sufficient for your workflow. DAWs offer editing features and the ability to create scores and parts from the Musical Instrument Digital Interface (MIDI) sequences created within the application. Any edits you make to MIDI and instrument tracks are immediately reflected in the notation. If you need to quickly generate a score of a song for musicians, the included notation features within your DAW will often be sufficient.

The differences between the DAW and a dedicated notation application lies in the flexibility and layout options. You may need to generate different parts from a score, each with its own font size. This may not be possible within the DAW application. Notation software is aware of the instruments you are using, meaning that it will warn you if a note you enter is not playable by the instrument. Notation software can guide you when working with transposing instruments, that is, instruments that sound in a different key than written. For example, a B-flat clarinet is a transposing instrument because when you play middle C on the instrument, the sound coming out of the instrument is B-flat below middle C. Finally, dedicated notation software allows you to focus on the notation, and not all the other elements within a DAW.

DOI: 10.4324/9781003345138-11

Music Notation Types

Music notation can be presented in several different ways depending on the genre or style. Some genres describe notated music as "charts." This term is mainly used by popular and jazz musicians. Traditional musicians will refer to music as parts or scores. There are eight types of music notation that you should be familiar with. Each type has its own layout and style.

Chord Charts

Chord charts are simple in that they only contain the chords, meter, and the form of the song. Chord charts are appropriate in situations where musical parts are improvised or already known, but the form and chords may need to be referenced. They are also helpful for a musician who needs a simple guide that they can improvise over. Chord charts do not contain the melody or any specific instrumental parts.

Sheet Music

This is the most common type for of popular music and songs. Sheet music contains the piano part, chords, lyrics, melody, and form. The piano is the only instrument in the music, although some sheet music has versions for other instruments such as a guitar. The music is a representation of a popular song and may not follow the exact form or style of the audio recording.

Songbooks

A songbook is a compilation of songs resembling sheet music. Songbooks come in different styles and for different instruments. They can be a collection of songs from an album, soundtrack for a film, or a musical. The instrumentation is often for piano and voice. Like sheet music, the chords and form are included. Sometimes, the songs are shortened to save space.

Lead Sheets

Lead sheets are popular with songwriters needing to copyright their songs. Lead sheets contain the chords, lyrics, and melody lines of the song. Instrumental parts, such as one for a piano, are not included. Lead sheets are the simplest way to present a song.

Fake Books

A fake book is a large book of music that contains only the melody line, lyrics, and chords. Fake books are popular with jazz musicians. These books are designed to present a large amount of music with the greatest amount of flexibility. The sections of the music are indicated, but the form is left up to the musicians. The piano, guitar, and bass parts must be improvised. Fake books often use a slanted notation font. Fake books are essential for working jazz musicians who must be prepared to play a variety of music.

Master Rhythm Charts

Master rhythm charts are often used by drummers who need a quick reference for a song. The main rhythms are indicated along with the form of the song. Sometimes, important lyrics or melodic lines are added for reference so that the drummer knows where they are in the song. Chords are included in case a bass player is reading from the same part as the drummer. Dynamic markings indicating how loud or quiet sections are sometimes included.

Music Score

A music score contains all the music for all the instruments required for the composition. Scores are common with music written for orchestras, choirs, or other ensembles. A score can also be music written for piano and voice or a small ensemble like a string quartet. The score contains all the elements for each instrument, including dynamics and articulations. Scores are a road map for conductors, who must lead the ensemble through the entire piece. Scores are notated so that the conductor knows the sounding pitch of all instruments. Scores are often printed on large pages to accommodate all the instruments listed on the score.

Music Parts

Parts are related to the score. Parts contain the music for each instrument listed in the score. Parts are notated for the written pitch of each instrument. A part for a B-flat clarinet will be written so that the clarinetist knows what notes to play. The layout for parts sometimes includes larger margins so that musicians can add their own notes to the music. Music that requires several pages will be laid out so that page turns happen when the musician is not playing.

Music Copyist

A music copyist prepares written music for performance or print. A copyists must present music scores and parts in a clean document that is easy to read. A copyist must also work quickly to make corrections and changes and meet the deadline of a client. The value of a copyist depends on two variables: accuracy and speed. Accuracy is critical because a score that requires multiple corrections after a review from the client means that the client is losing time and the copyist money. Speed is important because clients often work with strict deadlines and need jobs completed quickly. A music copyist must find a balance between these two variables to be effective and useful to a client.

A music copyist must deliver notated scores and parts for a variety of situations. Each situation has distinct requirements for the notated parts. Scores must fit all the instruments on a single page but still be easy to read. Parts for individual musicians should use as few pages as possible. If a page turn is required, the page turn should be at a point where the musician is not playing. Lyrics must be spaced such that they are easy to read and correctly placed underneath the notes. A copyist must be able to adapt to the situation and deliver quality results.

A singer might hire a copyist to generate charts for several songs they want to perform with instrumentalists or other vocalists. A music copyist must know how to create charts for each type of musician in the band. A singer may have a piece of music that they want transposed to another key that is easier for them to sing. A copyist must know how to transpose music. A singer might give a copyist a MIDI file of the song they want to learn the copyist need to convert the MIDI file into notated music and then add lyrics to the score.

Composers will often need a score of the music they have written within a DAW. This is often true with film composers. In this situation, the copyist must import the MIDI file exported by the DAW and create a score with the correct instruments in the correct order and the correct notation. Transposing instruments must appear accurately in the score and in the individual parts. The composer will often make changes to the score once rehearsal begins, and the copyist must quickly make corrections and hand out new parts.

Transcribing music is another task of a music copyist. A musician may need an audio recording converted into notation. Another musician may want a piece of music written for one instrument transcribed to a different instrument. Finally, a songwriter may have written a song in a DAW and wish to convert all the information into notated parts to hand out to a band. A copyist must be able to deliver accurately and quickly whatever the client wants.

The goal of a copyist is to deliver music scores and parts that are accurate and not in need of corrections. The copyist must be an effective proofreader and catch errors before sending the material to the client. A copyist is detail oriented and must compare the original score to the one that was created and ensure that every note and articulation is the same. Sometimes, a copyist catches errors on the original that must be clarified with the client before the final scores and parts are generated.

I have worked as a music copyist in the past and found it to be an exciting and challenging way of earning an income. I met a variety of musicians and was able to work on some interesting and exciting productions such as scores for films and music for books. It was also a learning experience since there were often requests that I was not familiar with but quickly had to figure out. Copyists can charge by the hour or page. Most clients prefer to pay by the page, so it is up to the copyist to determine a fair rate for both parties.

Notation Software History

Early notation software was difficult to use because of the operating systems and the graphical user interfaces that were used at the time. One of the earliest notation programs was a program called SCORE. The application was developed by a professor at Stanford University, Leland Smith, during the 1980s. The software ran on PC computers running Microsoft DOS, a command line operating system that offered no graphical user interface or mouse support. The entire operating system ran from a command line, meaning that everything that you wanted to do needed to be typed into the computer. If you wanted to see a list of files in a folder, you typed the command "dir," which would then generate a list of the files on your screen.

SCORE was an extremely flexible and powerful notation software but very difficult to use. Note entry into the application required you to enter multiple lines of code. This meant that you had to be extremely organized and careful when entering scores. Each line of code specified a particular aspect of the notation, for example, the pitches of the notes. Another line of code specified the duration of those notes, while another line determined any articulations and slurs. Once all the data was entered, you then waited for the application to render the score for display on your screen. The rendered score on the computer screen was crude and rudimentary due to the operating system. You needed to print the score using a specific printer protocol called PostScript to see the final product. The process was time-consuming, but there are many who will argue that SCORE generated the best-looking scores, even by today's standards.

As the Windows and Mac operating systems developed, notation software became easier to use and offered more features. Today you have several options for notation software, each of them being very good and offering several options. Your choice of which notation software to use depends on your own experience, the types of scores you wish to create, and probably more importantly, the amount of money that you want to spend. If you need to quickly generate parts from your existing sequences, it is worth spending some time learning the features of the notation options within the DAW that you already are using. You might find that it is sufficient to meet your needs at the time. If you need more features, many companies offer time-limited demo versions for you to try out their software.

Notation Process

When looking at professional notation software, the first step is understanding how the notation process works. The notation process is the same whether you are creating a score by hand on paper or using software. There are four stages involved when creating a music score: score setup, note entry, layout, and delivery. Each application follows the same type of process but differs in the execution of their process.

Score Setup

The setup process allows you to define the instruments, the clefs, the key signature, and time signature. Some applications will create a limited number of measures for you while others require you to specify a number. You always have the option to add or remove measures. Most applications will automatically assign sounds based on the instruments you choose. Notation applications include a sample library of sounds consisting of orchestral, band, world, popular, and keyboard instruments. If you have a particular sample library you are comfortable with, some applications let you choose your own library. I recommend this for advanced users since there are often several steps to loading your own libraries. The included libraries may not sound as good as other samples but remember that you are using the sounds for reference. The included libraries are closely integrated into the software to offer optimal performance.

The setup process usually involves choosing the paper size used for printing. You can always change the paper size during the layout and delivery stages, but it is helpful during the setup stage because the software will automatically space the lines of music and measures to optimally fit on the page. Notation software will display the score on your screen exactly as it will print. This is a significant advantage over earlier applications like SCORE.

All the notation applications use custom font packages to create the notes, text, and graphical elements on the score. Each manufacturer develops their own fonts to produce professional scores. Some applications offer different fonts depending on the type of music you are creating. Jazz scores use a different font than orchestral scores, so notation software will include a custom font for jazz. These font packages assign computer keyboard characters to note names and symbols. These fonts are designed specifically for music and offer high-resolution printouts. One of the major advantages of using notation software for creating scores and parts is that you can achieve excellent results that musicians will be able to read easily.

The setup process will ask for the composition title, subtitle, composer, year, and copyright information. You can add other elements such as page numbers and headers or footers on each page. All of these elements can be changed at any time during the notation process. If you are not sure of the title of the piece, you can enter a placeholder and change it later. All this information helps the software create a working layout for you to use while entering the notes. Once the initial setup is complete, you can begin entering notes.

Note Entry

Note entry is the most important stage of the notation process. This is where you specify the pitches and durations for all the notes on your score. This is also the time to enter items such as slurs, dynamics, and articulations. If you are working on a song, then this is where you would enter the lyrics and any chord diagrams. Most notation programs offer multiple methods of entering notes into your score. How each program executes each method can differ, but the fundamental steps are similar.

I will point out that notation software and note entry is not an automated process. All notation applications assume that you know how to read music and understand durations, articulations, and symbols. The software will not make decisions for you; you must tell it where you want the elements placed on the score. What the software does is automatically adjust the spacing of all the elements so that the score is easy to read, and the page does not look cluttered.

Computer Keyboard

The most common way of entering notes into your score is through manual entry. You can use a mouse to position the location and click to enter the note. This method is slow and not recommended for larger projects. You can also use the computer keyboard to type the notes, as if you were typing an email. The method for defining the pitch and duration of the note differs between programs, but most allow you to specify both elements from the computer keyboard. Once you are comfortable with this process you can quickly enter notes into a score as easily as you type an email. Many programs offer shortcuts, enabling you to speed up the process.

MIDI Keyboard

Using a MIDI keyboard to specify the pitch of a note is another option. If you are comfortable using a MIDI keyboard, then will find this method very efficient. In this scenario, the computer keyboard is used to determine the note duration and then you play the pitches on the MIDI controller. The advantage of using a MIDI controller is that you can easily specify the octave of the note you are entering. You can enter chords all at once, which you cannot do with a computer keyboard. If you are not comfortable using a MIDI keyboard, this method could be slower than manually typing with a computer keyboard.

Real Time

Some applications give you the option of playing the notes in real time. This is the same as recording MIDI information into your DAW. You simply specify the tempo, the number of count-in measures, and then press record. A click track will help you keep time as you play. The software will notate the pitches and durations as you play them. As with a DAW, the application allows you to quantize notes on entry or afterward so that any timing imperfections are corrected. If you are comfortable playing a MIDI keyboard, this is an efficient way of entering notes into your score.

Import

Most notation programs give you the option of importing data and converting it into notation. The first thing to understand is that all notation software programs use a proprietary file format to save documents. This means that the notation created on product A cannot be opened in product B. Each application has its own way of how the information is stored and thus saves the file using a proprietary format and file extension. However, this does not mean that we cannot export data from one notation package and import it into another.

MIDI Import

The first method of importing data is using MIDI. You can export the MIDI information from your DAW or other notation application as a standard MIDI file. The MIDI file can then be imported into the notation application. Keep in mind that the information will import the MIDI data exactly as it was recorded; any timing inaccuracies will be imported as well. If you are exporting from a DAW, it is a good idea to quantize all note start times and durations to ensure a smooth export and import. Many composers who do mockup scores on their DAW will export the MIDI information into a notation program in order to create the score and parts needed for musicians to play.

However, it is important to remember that MIDI information is strictly note information. Exporting a MIDI file from one notation application will only export the note pitches and durations for each of the instruments used in the score. Information such as the key signature, time signature, and measures are included in the MIDI export. Information such as articulations, dynamics, chord symbols, and lyrics are not included in a MIDI export. This means you can get all the pitches from a song, but you will need to manually enter the lyrics in your own application.

MusicXML

To overcome some of the limitations of importing and exporting MIDI files, an interchangeable file format was developed called MusicXML. MusicXML is a special language that allows additional information besides note data to be imported and exported. MusicXML can import and export lyrics and in some cases articulation markings. If the music you are working with does not have lyrics or articulations, then MIDI files are sufficient for transferring information.

Audio

Importing audio into a notation application is handled by third-party applications. The third-party application attempts to convert audio information into MIDI. The clearer the audio file, the more accurate the conversion. Single melody lines are easier to work with than complex chords. The results can vary. Sometimes, the generated MIDI file requires more time to edit than it would to enter all the notes.

The good news is that many DAW applications now offer the ability to convert audio into MIDI. The results will vary, but you do not need to purchase a specific application if your DAW can accomplish the task. The DAW can convert the audio into a MIDI track which you can edit and correct any mistakes before exporting the music to a MIDI file.

Scanning

Another way of entering notes into your notation program is to use an optical character recognition (OCR) software package designed specifically for music. The quality of these applications varies, and the accuracy depends on the visual quality of the music being scanned. Handwritten scores are the most challenging and require the most time. These music OCR applications will scan a sheet of music and generate a MIDI or MusicXML file which then can be imported into your application. If you are scanning a handwritten score, keep in mind that you may end up with many errors. In some cases, it might be faster to create a new score and manually enter the notes rather than correcting errors.

Text Elements

After you enter all the notes into your score then you need to take care of all the text elements such as articulations, dynamics, expressions, lyrics, chord symbols, slurs, and other lines. You can enter all these elements during note entry or afterward. I find it much easier to complete all the note entry first and then go back and make another pass to enter all the text elements. How text elements are entered varies between applications. You might find some applications to be very intuitive while others are not. Many applications offer keyboard shortcuts to speed up the process. In most cases, you will have the option of entering these elements using a computer keyboard or your mouse.

Layout

The layout stage is where you make all the decisions on how the page looks. In this stage, you choose the margins, page size, and the size of the staff and notes. The application will default to a standard size suitable for printing an easy-to-read score. You may wish to make the music larger or smaller, depending on the situation. To change the size of your staff and notation, some applications use a scaling function allowing you to choose a percentage of the original size. Other applications have you choose a font size and then adjust all the other elements. Either method will automatically adjust all the elements so that they are spaced correctly.

The layout stage gives you a representation of what your page will look like when it is printed. Modern notation software packages try to make the graphical user interface representative of what the score will look like at all times. However, when working on the layout, you often have additional options to adjust things like the margins and immediately see what that will do to your score. Some applications will still allow you to move certain elements when you are working on the layout while others will restrict that and require you to go back to another view in order to adjust and correct the note entry.

Advanced options include changing the number of staves on a page, the number measures per staff, and the spacing of notes. This can be done globally or on specific measures or staves. Manually adjusting these elements can lead to problems if you are not careful. There might be situations where custom adjustments are necessary, depending on the score. However, I recommend that you leave that decision to the software unless you have very specific needs for your score.

Delivery

The delivery stage is where you prepare the finished score for printing or export to a specific file type. The delivery stage gives you a preview of your finished score so that you see what it will look like once you print or export an image of the score.

Printing

Printing a score is the final stage of generating music for musicians to read. Notation applications will often offer you a print preview so that you can see your score as individual pages on your screen. Your printing options will depend on the type of printer connected to your computer. You also have the option of printing your score as a PDF file. If you are working on a score with multiple parts, then this is the stage where you can choose whether you are printing the score and/or the individual parts. Once again, the ease with which you can move between these various decisions depends on the application.

Exporting

Applications differ in their export options. MIDI and MusicXML export options are included with all notation applications. Some applications add the option to create an audio file of your score, usually an MP3 or WAV file. The audio export uses the sound library currently loaded with your score. When exporting audio, you may want to open the mixer for the application and adjust levels to ensure that all parts are heard.

Other export options of image files can be added to a document or a web page. Your choice of image files depends on the application. Some allow you to export the image as a vector file like a Tagged-Image File Format, which can be scaled to multiple sizes without affecting the clarity of the image. Others also allow you to use a compressed image file like a JPG. Always check the documentation to verify export options.

Summary

In this chapter, we looked at the notation process and computer music notation. We defined the role of a music copyist and the responsibilities of the copyists. We looked at the general features of music notation applications and how we can enter information into the application. We examined different note-entry methods depending on your skills and resources. We learned that importing and exporting information is an important part of the music notation workflow.

Computer music notation is not a task that everyone will enter. I added this topic to the book because I wanted you to be aware that this option exists. There are some of you that will find this information useful. If you plan on working with sheet music and other musicians, then it is worth your time to investigate notation software packages and determine which one will suit your needs and budget. I recommend taking advantage of time-limited demos to test various programs.

For many of you, the notation features offered by your DAW will more than meet your needs. If you need to quickly generate a printed sheet of music to go over a song with someone and make sure you are both playing the same chords, then the features within your DAW might be enough to get that job done. In the end, the decision to use computer music notation software is entirely up to you and your specific workflow.

Activity 10

Computer Music Notation

In this activity, we will look at notation software and if you are feeling adventurous, try entering some notes.

Dorcio SE

There are several free notation programs available, all with varying degrees of features. In this activity, I will be working with Dorcio SE by Steinberg. I am using Dorico SE because I find the application intuitive in its layout and note entry. As we go through the application, you will see that the notation process it uses closely matches the process I presented in this chapter.

Dorico SE is free, but you do need to create an account with Steinberg to download and activate the application. Start by going to the Dorico SE page: https://www.steinberg.net/dorico/se/. On the center of the page, click on the Download For Free button and follow the instructions. Once you create your account, you need to download and install the Steinberg Download Assistant. The Download Assistant will also install the Steinberg Activation Manager to activate your software. Follow the prompts during the process to complete the installation and activation of Dorico SE.

I have included a simple arrangement of the Minuet in G by J.S. Bach on the companion website. I created this score in Dorico SE. If you decide to follow the instructions below, you will create a score that looks exactly like the one I have provided.

1 Launch Dorico
2 Go to File – New
3 Click on Add Single Player

   a From the left column choose Keyboards
   b In the next column choose Piano
   c Click Add

4 At this point, it would be a good idea to save our project

   a Go to File – Save
   b Save your project

5 Click on the Write tab
6 Expand the area on the right side

7 Click on the two sharps symbols to open the Key Signatures panel

a Choose the Major key with one sharp
b Click on the symbol and let go
c Move your mouse over to the first measure
d Notice that your cursor is a key signature
e Click in the first measure to set the key signature

8 Click on the 3/4 button in the right panel

a Select the 3/4 time signature from the list
b Click in the first measure to set the time signature

9 Click on the Bars button in the right panel

a At the top of the panel under Insert Bars change the number 1 to 15
b Click on Insert Bars
a Now your piece has 16 measures

10 In the left panel, select the quarter note

b Mouse over the staff in the first measure until you see the outline of notes
c When you find the D click on it
d A quarter note D is entered

11 Now select an eight note from the right side

a Mouse over to the second beat and find G and click
b Repeat the process for the notes A, B, and C

12 To exit Note Entry, press Escape twice

13 If you are feeling adventurous, you can try a faster method of note entry

14 Each note length is assigned a number and you can use the keypad or the keyboard to enter the numbers

a To select a quarter note, press the number 6
b To select an eighth note, press the number 5
c To select a half note, press the number 7
d To add a dot onto any note, press the period on the keyboard

15 To determine the pitch, you press the letters A through G

16 Manually enter the D in the second measure of the first line

a The next two notes are quarter note Gs
b Press the letter G

- c The note is in the wrong octave
- d To lower the octave, press Ctrl/Cmd + Alt/Opt + Down Arrow
- e To go up an octave, press Ctrl/Cmd + Alt/Opt + Up Arrow
- f To go down a step, press Alt/Opt + Down Arrow
- g To go up a step, press Alt/Opt + Up Arrow

17 It takes some practice, but it is faster in the long run

18 Dorico takes care of all the spacing for you, so you do not need to concern yourself with that

19 When you are finished entering the notes, you need to add the titles and the composer

20 Click on File – Project Info

- a Click on the Project tab on the left
- b For the Title, enter Notebook for Anna Magdalena
- c For the Composer, enter J.S. Bach
- d Now click on the Flow 1 tab
- e For the Title, enter Minuet in G
- f Click Apply and then Close

21 The number 1 in front of Minuet is unnecessary

- a Double-click on the number 1
- b Some strange text appears
- c Delete the following text: {@flowNumber@}.
- d Click on a blank section of the score

22 Now your score is complete

23 If you want to print or export your score, click on the Print tab

- a Choose a Printer to print a physical copy on paper
- b Choose Graphics if you want to export to a PDF or other graphics file

24 Save your project and quit Dorico

This activity is an introduction to notating music with Dorico. If you want to explore notation more, you can visit the Steinberg website to find tutorials and learn more about Dorico.

Terms

Chart Another name for a piece of sheet music.

Chord Chart A music score that only shows the chords, meter, and form of the song.

Copyist A music professional responsible for creating music scores, parts, and other notated music.

Fake Book A large book of music that shows only the melody line, lyrics, and chords.

Lead Sheet A music score that only shows the chords, lyrics, and melody line of the song.

Master Rhythm Chart A music score that shows the main rhythms for a song and the form.

Music XML A special file format that stores notation and lyric information. This format is useful when sharing a notation project between different applications.

Optical Character Recognition OCR A feature included in some scanners that can read text and music notation from a scanned page.

Score Music notation that shows all the instruments used in a musical composition.

Sheet Music A generic term for a piece of notated music.

Songbook A book that contains a collection of songs.

Chapter 11

Growth and Development

Putting It All Together

Some of you may have been gathering equipment while reading this book, others of you may be starting to put together your list. In either case, we should spend a little time talking about putting together your home studio. I want you to keep in mind a few things as you go through this process.

First, keep in mind that your studio is a constantly evolving system. You may settle on equipment to get you started, but do not be surprised if later you wish to upgrade or add to your studio. If you are just starting out, you cannot anticipate what your needs will be in a year. As you develop your skills, you will probably discover new ideas that you wish to try out. For example, you might be inclined to purchase a hardware synthesizer so that you can experiment with designing your own sounds. You may find yourself working with other musicians and need a way to record multiple instruments at the same time. You may also decide to take your productions on the road and perform in front of people. These are all possible scenarios.

If you are unsure as to the equipment that you need when starting out because you are not certain about the type of music you will be producing, then I suggest building the basic set up that I recommended at the beginning of this book. This configuration has all the equipment you need to work on a variety of different music styles. Below is the recommended system.

- Desktop or laptop computer
- Audio interface with two inputs into outputs
- A dynamic or condenser microphone
- A 25- or 49-key MIDI controller depending on your keyboard skills
- A pair of headphones and/or speakers

Many manufacturers include an entry-level digital audio workstation (DAW) application with their audio interface. This might be an intro version of applications like Ableton Live or Steinberg Cubase. These intro versions are usually limited to eight tracks, a small collection of audio effects,

DOI: 10.4324/9781003345138-12

and a couple of virtual instruments. These versions are designed to get you started with making music, and there are usually tutorials and templates to help you get started. Depending on your skills, you may find yourself outgrowing these applications and needing to upgrade to a full version. You may also discover that the included application with your audio interface does not meet your creative needs. In this case, I recommend downloading trial versions of other DAW applications until you find one that has a desired workflow.

Early on in this book, I mentioned that there is no single solution when it comes to DAW applications. Each application has its strengths and weaknesses. Over time, you may discover that you will need to run multiple DAW applications to accomplish different tasks. You may use one application when working exclusively with audio and needing to do many complex edits. Another situation might have you creating remixes of music in which case a different DAW application might be better suited for the task. Once you understand the functionality and workflow of a DAW application, learning a second or third application often takes less time. This is because you are already familiar with how an application functions and simply need to learn how the second application handles similar tasks.

You need to be comfortable with the fact that your studio will require additional investments over time. As technology evolves you may want to take advantage of new developments, or you may want to expand the type of work you can do. As your skills develop, there is a chance that you will outgrow the technology that you currently use. In the next section, we will look at different scenarios that might compel you to upgrade and add to your studio.

Upgrading

As we come to the end of this book, I want to talk about upgrading. No matter how well you plan and design your home studio, you will eventually want or need to upgrade your software and technology. In fact, you might go through several upgrades until you reach a point where you feel you have exactly what you need. How often you perform upgrades depends on your needs at the time, the quality of your equipment, and how up to date you want or need to be. Let us look at some examples and variables to consider.

DAW Applications

Some DAW application manufacturers operate on yearly subscriptions. During that year, you are entitled to feature updates and other benefits. When the subscription expires, you all lose access to the application. To continue to use the application, you need to renew your subscription every year. This is an expense you can budget, and if you depend on the application, set your subscription to auto-renew so you do not have to remember to renew. With a subscription service, you always have access to the latest version of the software.

Other companies allow you to purchase a permanent license for the application. The license will grant you access to the application, and you can run it as long as your computer and operating system supports it. When a new version is released, you can opt to upgrade to the new version for a nominal cost, or you can continue to use the older version. Keep in mind that when a new version is released, most companies will cease making updates for the older version. If you upgrade your computer operating system to a version that does not support the application, then you will need to purchase an upgrade for a version that is supported.

A few companies allow you to purchase a permanent license and receive updates for a year. After the first year, you can still run the software, but you do not receive any updates. To receive updates, you need to purchase a service or upgrade plan, which covers you for a year. This feels a bit like a subscription, except if you choose not to purchase a plan, you can still use the current version of the software until it is no longer supported by your computer or operating system.

I keep mentioning the operating system and computer as a variable when deciding on upgrading your application. Software developers must consider changes in operating systems when developing their products. Sometimes, an operating system update is significant enough that a manufacturer must rewrite the software code for large sections of the application. When they finish adapting the application to the new operating system, they find themselves supporting two versions of their product: one for the older operating system and the other for the newer. The company may not have the resources to support both versions for very long, which means at some point they announce that they will no longer offer updates for older operating systems.

If you choose not to update your operating system, then you can continue to run the older version of the application. Many users choose this option, only upgrading when their computer stops working. If the application does what you need it to do, then upgrading to a newer version may not be compelling. Weigh your options carefully and consider your needs. Some of us look forward to upgrades because of new features and tools; however, if those new features and tools are not necessary for your workflow, then maybe the upgrade is not required.

Computer Hardware

We would like our computers to last forever, but at some point, they will fail and need to be replaced. A Windows PC desktop computer provides some flexibility. If a storage drive or graphics card fails, you can replace that yourself. Sometimes RAM will fail, which can be replaced by the user. If the motherboard or CPU fails, the ability to replace those components depends on the age of the computer. CPU manufacturers introduce new CPUs every 12–18 months. Sometimes these new models are backward compatible with existing motherboards, other times an updated motherboard is required. If either your motherboard or CPU fails, the replacement must be compatible with the existing component. If not, then you must purchase a new CPU and motherboard. The graphics card and storage devices you have will most likely work with the new configuration, the RAM may not.

Windows PC laptop computers vary on what can be repaired. Some laptops are designed to be serviced by the user; others are not. Components do cost more for laptops, and storage drives and memory are usually the only two components you can replace. The CPU and graphics card are integrated into the motherboard; if any one component fails, the entire motherboard must be replaced. Screens can fail on laptops, which can be costly to repair. If your laptop is your only computer, then an extended warranty is worth the cost.

Since Apple computers are closed systems, you must take the computer, desktop, or laptop to a certified service center for repair. If the machine is still under warranty, then the repairs are usually little to no cost. Systems that are out of warranty will cost significantly more to repair. Apple offers extended warranties on their desktops and laptops. Purchasing the warranty provides you with some insurance in case anything goes wrong with your computer.

Given how much we rely on our computers for music production, it is worth considering replacing your computer every four to six years, depending on how much the computer is used. Replacing a computer before the older one fails helps prevent downtime and stress. I know several users who were in the middle of a job when their computer failed and had to scramble to purchase a new one to meet the deadline. The older computer can serve as a backup if needed. How often you replace a computer varies and depends on your personal desire to always run the latest operating system and applications.

Audio Interfaces

As with DAW applications, a manufacturer may decide to stop supporting older audio interfaces by not updating device drivers to support newer operating systems. If a new operating system requires the manufacturer to write new drivers, they may focus on supporting newer devices and drop support of older ones. Some manufacturers support older hardware for a lengthy period, others may not. If you registered your audio interface with the manufacturer, then you are probably on their email list so you will be notified when hardware is no longer supported.

Most of us end up upgrading our audio interfaces because we need more inputs or outputs. Another reason is that we want additional features such as on-board effects processing. Some audio interfaces have a digital signal processor (DSP) that will process specific audio effects. This feature may allow you to record with audio effects in real time. Other features such as multiple headphone connections and digital expansion are compelling reasons to upgrade an audio interface. When considering upgrading your interface, spend time reading critical reviews of the interface. You can also visit the manufacturer's website and download the owner's manual. The manual can often answer questions about features and functionality.

Speakers and Headphones

Speaker and headphone upgrades are the most common I have seen and experienced. It takes time to find a pair of speakers that work well in your room and provide you with the fidelity you want. If budget dictated your first pair of speakers, then you will be compelled to purchase a different pair when you have the funds. Entry-level speakers will allow you to work successfully, but as your ear develops, you may find the speakers lacking in detail and clarity. This can lead you through a series of trials and errors until you find a set that works for you. A pair of speakers may sound excellent in the showroom, but once they are sitting in your room and desk, you may feel otherwise. As I stated in Chapter 7, speaker decisions are subjective and spending more money does not mean better quality.

Headphones are subjective as well since you need to find a balance between sound quality and comfort. I have a pair of headphones that have excellent sound quality, but their lack of comfort prevents me from using the headphones for a long period of time. Over the years, I have settled on two different models of headphones and alternate between the two. Headphones do wear out, and you may hear crackling as the diaphragms break down. This is why I keep a second pair of each model in my studio.

Microphones

I think it is safe to say that we never upgrade our microphones, we simply add more to our collection. Microphones are always useful to have and even the first microphone we purchased will find a use in later projects. If you are a drummer, you may be looking at a set of drum microphones. Instrumentalists might look for microphones better suited for acoustic instruments, usually purchased in pairs. Vocalists may want a large diaphragm condenser microphone specifically designed for the voice. Microphone choices seem limitless and deciding on your next microphone may be difficult. Your budget may help narrow your choices, but you will discover there are many options for every budget. Like every other purchase you make, do your research, and read reviews. Take advantage of any opportunity to test a microphone before purchasing.

Plug-Ins

There will be a point where you will want to add to your plug-in collection. There might be a synthesizer or effect that is not included with your DAW that you want to work with. You may also need certain plug-ins to collaborate with others. The plug-in market is vast with countless options. Some plug-ins emulate existing hardware synthesizers and effects while others are entirely new creations.

The plug-in market is competitive and with that comes the convenience that nearly all manufacturers offer time-limited trials. This is the best way to evaluate a plug-in and determine that it offers the features you want and is compatible with your DAW application. We covered plug-in formats in Chapter 4, and you will recall that your DAW application will support VST, AU, or AAX. While most plug-ins support all three formats, always check the specifications. The trial version is the best way to check for compatibility.

In my experience, more money is spent on plug-ins than on hardware. There are many appealing instruments and effects with attractive pricing. It is very easy to find yourself buying plug-ins as soon as they are announced just because it is something that you think you might use in the future. You may find yourself with dozens of unused plug-ins on your system. Be selective and take advantage of the trial version. This is the only way to determine whether the plug-in is useful to you.

Instruments

Hardware synthesizers, drum machines, guitars, basses, and other instruments may become part of your studio depending on your workflow and productions. Recommendations are beyond the scope of this book, but you can apply the same decision-making process to other hardware as you have with your DAW hardware and software. I do recommend visiting a store to try the instrument in person. Video reviews are helpful and informative, but they cannot replace the experience of placing your hands on the instrument. Quality instruments require a significant investment, so once again, do your research and ask questions.

Your reasons for wanting to upgrade or add to your studio are unique. The examples I have listed above are situations that other users and I have experienced. Many of you may not find a compelling reason to upgrade your existing studio and that is fine as well. I do want you to be aware that the possibility exists and that it should not surprise you. If you are committed to your music productions, you will find yourself wanting to grow and expand your skills and abilities.

Budgeting

I often tell new users that they picked an expensive hobby when they first entered the world of music production. As affordable and accessible as technology has made it for us to build our own home studios, there is still an investment of finances and time. The cost is significantly less than it was 20 years ago but there is still a cost. Even if you are making money from music productions, you still must consider your budget when planning future upgrades and enhancements.

In the previous section, I recommended that you plan on upgrading your computer every four to six years, depending on its use. This is an expense that you should try to plan and budget for so that it does not catch you off guard when the time comes. Planning is important because there is a good chance that your next computer will cost more than your first one. You may decide to purchase a machine with additional RAM, enhanced graphics, and additional storage. If you are a laptop user, you may decide to purchase a new laptop with a larger screen. Desktop users may want to have a system that supports two screens. You may be working on projects that have a significant number of audio and instrument tracks thus requiring a more powerful CPU.

This is also true of audio interfaces. When you decide to upgrade your audio interface it will most likely be significantly more than your first one. Higher-end audio interfaces do offer better fidelity and increased audio processing. If you currently have an eight-input entry-level audio interface, your next audio interface will easily cost more because you still need the eight inputs and may want to take advantage of advanced features such as DSP processing. I cannot recommend the timeline as to when and if you should upgrade your audio interface. Your needs will dictate when that moment comes. But you do need to be aware that it probably will happen.

Adding microphones to your studio can be costly. You may have purchased an entry-level microphone kit for drums but after upgrading your audio interface you may be compelled to purchase new microphones for your drums. Depending on the model you choose, this can be a significant expense. You may end up needing to upgrade a couple of microphones at a time depending on your budget. As I stated earlier, your options for microphones are as vast as is the price range. My recommendation when budgeting for upgrades is to create a priority list. Even with an entry-level microphone kit for your drums, upgrading your audio interface will have a significant impact on how those microphone sound. Therefore, you may not need to upgrade the microphones immediately.

I know many users who set aside money every month into a savings account or some other account that they do not touch. This method of budgeting is helpful because you can build the funds you need for your upgrades over time. Many users find this helpful because they can plan for their upgrades. This also prevents having to spend funds all at once. How you budget for your upgrades is entirely up to you, but I do recommend coming up with a plan for upgrades. As pleased as you are with the equipment you are currently using, there is a very good chance that you will be compelled to upgrade parts of your studio in the future.

How Much Is Enough

This discussion of upgrades can lead to the question of how much equipment you really need. Trade publications and internet advertisements showcase dozens of manufacturers trying to convince you that their products will enable you to make high-quality music. They also may claim that their technologies will allow you to work faster and create better music. These advertisements are difficult to resist because the demonstrations in the videos sound amazing. This is especially true of plug-ins.

As I stated earlier, the issue with plug-ins is that they are often competitively priced, often so low that purchasing them seems like a good idea. Even if the plug-in does not deliver everything it promised, the investment was low enough for you to forgive its shortcomings. This makes adding plug-ins to your system easy and tempting. After some time, you may find yourself with a dozen different reverb plug-ins. At this point, it is fair to ask, how many plug-ins do you really need?

Hardware products, such as audio interfaces, tend to remain relatively static. You tend to only upgrade hardware when it fails or if you want more features. High-end audio interfaces are designed for continuous studio use, so those purchases will last you for a long period of time, provided the manufacturer continues to develop device drivers for newer operating systems. The same is true for your DAW application. Once you commit to a DAW application, you will upgrade as needed. Some manufacturers offer small upgrades every year, while others will wait several years before releasing an upgrade. These investments are static because they do not occur often. Plug-ins are far more fluid, and because of low pricing, can lead you down a rabbit hole of constantly buying plug-ins.

Returning to the question of how many reverb plug-ins do we really need, it all depends on the type of work you do. There are many users who are quite satisfied using two or three different reverb types. You may find a certain type of reverb very good for vocals and a different reverb good for instruments. You may also discover that you have a single plug-in that works well with multiple sources and situations. A reverb plug-in may be based on a piece of vintage hardware that attempts to recreate the sonic characteristics of that hardware. Plug-ins based on plate reverberation units are very popular in plug-in format. If you are not careful, however, you may find yourself with four or five different plate reverb plug-ins that share similar sound characteristics.

The easiest way to avoid falling down the plug-in rabbit hole is to explore all the plug-ins that are included with your current DAW application. Manufacturers have gotten very good at offering a wide range of plug-ins as standard with the software. You may discover that the reverb plug-ins included with your software are excellent and provide a sound that meets the needs of your music productions. This applies to other effects such as equalization, dynamics, choruses, and delays. It is worth your time to familiarize yourself with the included plug-ins in a DAW application. If after some time you find yourself wanting something different, then download some trial plug-ins and listen to what they offer. There are many excellent offerings in the market, but you do not need to buy them all at once.

Closing Thoughts

In this book, we have looked at the basic terms, concepts, and technologies involved with audio production. My hope is that you have a better grasp of these items and that you feel more confident about the decisions you make when building your home studio. The goal of this book has been to provide you with essential information so that you can make informed decisions about the hardware and software that you purchase for your music production. I hope that you feel you possess the knowledge base needed to examine claims made by manufacturers when they state that their plug-in or application will fulfill all your music production needs. I have offered you recommendations and variables to consider when choosing hardware and software. Remember, there is no single solution that will meet everyone's needs; therefore, it is critical that you determine your own goals with music production before investing in hardware and software.

As your experience grows, you may find yourself wanting to learn more about newer and other technologies or techniques. On the companion website, I offer you a list of suggested readings based on what is currently available in the market. Some of the books I mention may be in the process of receiving a new edition. The books listed are ones I consider to be industry standards and used by many people as sources of information and reference. Some of you may choose to enroll in a program to increase your knowledge and experience.

For now, I do sincerely hope that you feel adequately equipped to begin building your home studio and begin your music productions. As new hardware products emerge, take the time to visit stores to try out the products and hear the results for yourself. Software and plug-in manufacturers tend to offer trials of their products. This is an excellent way to determine if a particular plug-in does what you needed to do. Your ears are your best tool to judge the effectiveness of these products. If the plug-in that you are trying does not perform as you expected, take the time to contact the manufacturer and see if they can provide you with some guidance. The best way to avoid purchasing something that you stop using after a few months is to try it out.

I have created this book to be an introduction to music technology but also to serve as a reference. You may not recall all the concepts and terminology mentioned in the pages of this book. It is for this reason that I have included in the index as well as a list of terms at the end of each chapter. Though the companion website will be updated as needed and will also serve as a reference for you. In closing, I wish you good luck and success in your journey into music production.

Index

For Product Safety Concerns and Information please contact our EU representative GPSR@taylorandfrancis.com
Taylor & Francis Verlag GmbH, Kaufingerstraße 24, 80331 München, Germany

www.ingramcontent.com/pod-product-compliance
Ingram Content Group UK Ltd.
Pitfield, Milton Keynes, MK11 3LW, UK
UKHW020008200926
475702UK00015B/2537
9781032384542